Writing Papers in the Biological Sciences

Writing Papers in the Biological Sciences

Victoria E. McMillan
Colgate University

A BEDFORD BOOK
St. Martin's Press New York

Library of Congress Catalog Card Number: 87-060597

Manufactured in the United States of America

7 6 5 4
k j i h

For information, write St. Martin's Press, Inc.
175 Fifth Avenue, New York, NY 10010
Editorial Offices: Bedford Books of St. Martin's Press
29 Winchester Street, Boston, MA 02116

ISBN: 0-312-89489-9

Bedford Books of St. Martin's Press

Publisher: Charles H. Christensen
Associate Publisher: Joan E. Feinberg
Managing Editor: Elizabeth M. Schaaf
Developmental Editor: Stephen A. Scipione
Production Editor: Chris Rutigliano

Copyeditor: Sally Stickney
Book Designer: George McLean
Cover Designer: Volney Croswell
Cover Photograph: Peter Menzel/Stock, Boston

Acknowledgment

Biological Abstracts, from Volume 83(3), Semiannual Cumulative Index and Abstracts Listings, February 1, 1987. Copyright © 1987 Biological Abstracts, Inc. Reprinted by permission of BioSciences Information Service (BIOSIS).

Preface

As a biologist who also teaches composition, I see many biology students who, although they have strong backgrounds in biology, have trouble understanding the writing tasks and challenges unique to their discipline. Biologists need to think creatively about science and they need to *write* about science clearly, accurately, and concisely. In fact, these two processes, thinking and writing, are as interdependent in science as they are in any other field. Unfortunately it is difficult to fit any extensive discussion of scientific writing into a regular semester-long biology course. I hope this book can substitute for such discussion, serving as a supplemental text for courses requiring laboratory reports or longer written assignments. For this reason, the book has been designed as a self-help manual, offering straightforward solutions to common problems and numerous examples of faulty and effective writing.

The book is intended mainly for undergraduate students who need a beginning knowledge of the aims and format of biological writing. It will also be useful for more experienced students, including master's and doctoral candidates who are writing longer papers and dissertations or preparing a first manuscript for publication. Chapter 1 focuses on the *biological research paper,* the primary form in which biologists communicate original findings. Students are expected to follow its style and organization when they write research projects or laboratory and field reports. Chapter 2 discusses the *biological review paper,* which resembles the library-oriented "term paper" assignments of advanced courses. Because both research and review papers must be carefully documented, detailed coverage of literature citations and biological formats for listing references is included in Chapter 4. Strategies for searching the literature and for drafting and revising are explained in Chapters 2 and 5 respectively, because many students are uncertain about how to carry out a written assignment. Chapter 3 discusses the design and execution of figures and tables, especially as they relate to the coherence of the rest of the text. Special problems of style, usage, and mechanics as they apply to biology, and special technicalities related to the preparation of the final draft are covered in Chapter 6. Chapter 7 offers advice on how to use writing effectively to record class notes, answer essay questions on exams, and prepare oral presentations. Finally, for readers wishing further information on specialized topics, there is an annotated list of additional readings.

Throughout the book I have tried to address the special demands and challenges of academic assignments in the context of how and why professional biologists write. I hope this handbook will convince students that scientific writing, including their own, need not be tedious and cumbersome, but should be clear, crisp, incisive, and engaging.

Acknowledgments. This book has been shaped, drafted, and revised with the assistance of many people. David A. Jolliffe, Department of English, University of Illinois at Chicago, served as consultant for the project and provided much valuable advice. The manuscript profited immensely from the criticism of the following reviewers: Richard F. Ambrose, Mitigation Program, University of California, Santa Barbara; Edward H. Burtt, Jr., Department of Biology, Ohio Wesleyan University; Elof A. Carlson, Department of Biochemistry, State University of New York at Stony Brook; John R. Diehl, Department of Animal Science, Clemson University; Malcolm Kiniry, Writing Programs, University of California, Los Angeles; Michael G. Moran, Department of English, University of Rhode Island; Bobette Vay Nelson, Department of Biology, University of California, Los Angeles; Carl W. Schaefer, Department of Ecology and Evolutionary Biology, University of Connecticut; and Josephine Koster Tarvers, Department of English, Rutgers University.

At Bedford Books, I also owe many thanks to Charles Christensen and Joan Feinberg for their continuing commitment to the book, and to Karen Henry for insightful comments about the first draft. I am grateful for the thoughtful recommendations of Elizabeth Schaaf and Nancy DeCubellis. Sally Stickney made many valuable suggestions for improving the clarity and consistency of the text, and Chris Rutigliano patiently guided me through the steps of converting manuscript into print. My largest debt is to Steve Scipione for his unfailing confidence and advice at all stages of the book's preparation.

Colleagues at Colgate University also helped in various ways. Rebecca Howard first saw the possibilities for my involvement with this project; her innovative development of the Interdisciplinary Writing Program at Colgate has contributed immensely to my own growth as a writer. I also thank Deborah Huerta for her help with the section on scientific literature and Ronald Hoham and Frederick Weyter for samples of their students' writing.

Finally, I owe special thanks to Robert Arnold, who not only critically reviewed several drafts of the manuscript, but also served as consultant for the chapter on figures and tables, for which he provided many of the examples. Without his encouragement and support, this book would not have been possible.

Contents

Writing Papers in the Biological Sciences

INTRODUCTION

How and Why Biologists Write

Science is a collective enterprise: its growth depends on the work and insight of many individuals and on the free exchange of data and ideas. Scientists must be effective writers because no experiment, however brilliant, can contribute to the existing fund of scientific knowledge unless it has been clearly and accurately described to other scientists.

Of course, many scientists write for the public in books, magazines, and newspapers intended for a general audience. Nevertheless, their professional work is published in scientific journals and aimed at a specialized group of readers, other scientists or nonscientists who have some background in the particular subject area. Biologists write two major types of papers for such journals: the research paper, or "scientific paper," and the review paper.

The *research paper* is a highly stylized report of original findings, generally divided into six distinctive sections: Abstract or Summary, Introduction, Materials and Methods, Results, Discussion, and Literature Cited. Although editorial requirements vary from journal to journal and authors must conform to these specifications, most scientific papers follow the same basic organization. Their format has been dictated by a long history of printing and publishing traditions that reflect the logic and elegance of the scientific method.

The format of a scientific paper reflects the scientists' obligation to make their assumptions clear, their methods repeatable, and their interpretations clearly separate from the data or results. Although a scientific paper may be descriptive, it is also a well-structured *argument* founded on supporting evidence. Scientific papers provide a forum for presenting one's own findings and conclusions and for arguing for or against competing hypotheses. In more practical terms, they also serve as a principal means by which a particular scientist's ideas and contributions are evaluated by his or her peers.

Academic assignments such as laboratory reports and independent research projects also incorporate the principles of scientific argument. They are excellent practice for writing a "real" scientific paper — which is the reason, of course, they are emphasized in biology courses. They force you not only to think like a scientist, but also to write like one. And most important, they show you that data cannot speak for themselves: you still have to incorporate them into a coherent, meaningful story, a scientific argument, that supports your conclusions.

The *review paper* is an article that synthesizes and interprets work on a particular subject area. Its format is not as standardized as that of a research paper (see p. 48). By bringing together the most pertinent findings of numerous papers from diverse journals, a review paper serves as a valuable summary of research. As a *secondary source,* it does not report the writer's original discoveries, but it does reflect his or her painstaking efforts to report accurately on the state of knowledge in a defined field. A good review not only synthesizes information; it also provides a critical overview of an important scientific problem.

Term papers or library research papers are the academic equivalent of review papers. In preparing them, you learn how to hunt down literature in a particular field, synthesize information, make comparisons and contrasts, and interpret and evaluate what you have read. Moreover, review papers immerse you in the *primary literature* of research papers, showing you how biologists think, write, and work.

CHAPTER 1

Writing a Research Paper

A research paper is a report of original research findings organized into several sections according to a format that reflects the logic of a scientific argument (see pp. 1–2). Professional scientists follow this format closely when they write a manuscript for publication. As a biology student, you will use the same organizational method for research projects and laboratory reports.

Many writers feel that the fairly rigid structure of a research paper actually makes it easier to write than a review paper. Once you understand what goes where, you can break up the writing into discrete tasks and tackle each one in turn. You do not have to compose the paper in any particular order (see pp. 91–92). Begin with the section that seems easiest; for many people this is Materials and Methods. The Discussion and Abstract are usually easiest to write last.

See Chapter 4 for instructions on how to prepare the Literature Cited section.

RECORDING AND ORGANIZING YOUR FINDINGS

■ Record data carefully and completely.

The quality of your data depends on many factors, including your experimental design and equipment, the kinds of information you choose to obtain, your skill and experience in carrying out the procedures, the statistical measures used to analyze your findings, and (for a field study) the extent to which environmental conditions make your observations feasible. Obviously you cannot foresee or compensate for all the problems that may detract from the validity of your results. However, you *can* begin the study prepared to record the data as completely and accurately as possible. Doing so will save you much work later when it comes time to analyze your findings.

Decide exactly what information you will need to record and in what form you will record it. Make up a preliminary data sheet, using headings to create a series of columns in which to enter the values you think you will need. Figure 1 is a sample data sheet for a study on the oviposition (egg-laying) behavior of a species of dragonfly. Its orderly arrangement allows you to record data quickly and systematically during the course of the observations, lessening the chance that you will forget a crucial bit of information. Notice that many of the column headings already include the units of measurement. The last column on the right-hand side provides space to record things that come up unexpectedly during the observation period.

A good data sheet is the product of much time and thought. Figuring out appropriate headings for all the columns forces you to evaluate the effectiveness of your methods and the rationale for collecting each set of information. However, even the most carefully prepared data sheet may prove to be unsatisfactory the first time you use it because methods or formats that seem feasible on paper do not always work in practice. You may need to modify your original sheet to account for problems you had not anticipated or for additional variables that need to be considered.

Here are a few more suggestions about recording results:

1. After each experiment or observation period, look over the data sheets carefully. Check that your writing is legible. Add additional comments or questions immediately, before you forget them. Do *not* recopy the data unless it is absolutely necessary. If you must do so, double-check everything to avoid introducing copy errors.

2. Wherever possible, incorporate the raw data into summary sheets as you go along. This is particularly important to do when you are recording many different kinds of information on the same data sheet. Because you will probably need to analyze certain data sets separately from the others, you can save time and effort by using a routine method for compiling related information. Also, summarizing the data as you go along allows you to be flexible and responsive to changing priorities during the research.

3. *Never* throw away any of your raw data sheets; you may need to look at them again later. Take the precaution of making duplicate copies of all rough data sheets, computations, summary tables, or other materials in case you lose the originals. Be sure to keep the duplicates in a different place.

Oviposition in *Plathemis lydia*

Date ____________ Sheet no. ____________

Field site ____________ Observation period ____________

♀ no.	Time ovip. ended (h)	Ovip. duration (s)	Pond site (marker no.)	No. other *P. lydia* at pond		Temp. (°C)	Other weather data	Comments
				♂	♀			

FIGURE 1. Sample data sheet.

■ Compile notes for each section of the paper as you do the research.

The bulk of your serious writing will have to wait until you have analyzed all the results. But if you try to visualize each step you take as part of a particular section of the paper, you will be better prepared to start writing once the time comes. Moreover, you will stay in touch with the aims and progress of your study and be able to modify your approach, if necessary, while there is still time to collect additional data.

One good organizational method is to have a separate file folder or notebook section for each part of the paper (Introduction, Materials and Methods, Results, and Discussion). While the study is still fresh in your mind, jot down procedural details, notes on the strengths and weaknesses of the experiments, and speculations about the significance of your findings. Record ideas, questions, and problems as they come to you and file everything in the appropriate place. Do not rely on memory. As good as yours may be, you are bound to forget something unless you write it down. The act of writing will impose order and logic on your activities, especially at the start of the project when you may be inefficient and disorganized. It will also help you crystallize your thoughts, gain perspective, and perhaps see new patterns in the data. You may find that a casual thought captured on paper eventually proves to be a flash of insight.

TITLE

The title identifies the important contents of the paper and orients the reader by specifying the writer's major findings or perspective. For this reason, abstracting and indexing services rely heavily on the titles of papers to organize large quantities of scientific information. Researchers also scan titles to pin down the most relevant references. A vague or inaccurate title can waste a reader's time by suggesting, erroneously, that the paper contains certain information. Even worse, a good paper burdened with a bad title may never catch the eye or the interest of many of its intended readers.

Many writers compose the title last, once they have written at least one draft of the paper and have a better understanding of their purpose and scope. Others devise a working title *before* they do much writing, to help them focus their ideas; then they revise it later, if necessary. See page 124 for instructions on preparing the title page of the final manuscript.

■ Make the title informative and specific.

Organize your title around important words (key words) in the study. Imagine that your paper appears in an index or abstract. How easily could it be located by someone looking up key words pertaining to your subject area? Does the title convey the major problem you set out to investigate?

The following titles, for example, are vague and uninformative compared to the revised versions following them.

VAGUE	Ecological Studies of Some Northern Lakes
SPECIFIC	Seasonal Algal Succession and Cultural Eutrophication in Three North Temperate Lakes
VAGUE	Dominance Behavior in Rhesus Monkeys
SPECIFIC	Androgen-induced Social Dominance in Infant Female Rhesus Monkeys
VAGUE	Aspects of Territoriality in Lizards
SPECIFIC	The Adaptive Significance of Territoriality in Iguanid Lizards
VAGUE	Some Changes During Menopause
SPECIFIC	Early Menopausal Changes in Bone Mass and Sex Steroids
VAGUE	A Look at Fungal Toxins
SPECIFIC	The Role of Fungal Toxins in Plant Disease

■ Be concise.

Make every word count. Omit unneeded "empty" words at the beginning of the title.

WORDY	Preliminary Studies on Primary Productivity and Phytoplankton Diversity in Four Wisconsin Lakes
CONCISE	Primary Productivity and . . .
WORDY	Notes on the Effect of Ambient Temperature and Metabolic Rate on the Activity of New Mexican Lizards
CONCISE	Effect of Ambient . . .

Replace wordy phrases with shorter phrases or single words.

WORDY	Why Felids Copulate So Much: One Possible Model for the Evolution of Copulation Frequency in the Felidae
CONCISE	An Evolutionary Model for High Copulation Frequency in the Felidae
WORDY	Studies on the Reproductive Biology of *Drosophila*, Including Sperm Transfer, Sperm Storage, and Sperm Utilization
CONCISE	Sperm Transfer, Storage, and Utilization in *Drosophila*

■ Include appropriate taxonomic information.

If your work features a particular species or larger taxonomic group, specify this clearly in the title. Species should be described by their full Latin names — both genus and species (see pp. 113–115). Include other information, if necessary, to orient the reader. Such information may include common names (if they exist) or the name of the family, order, or other important group to which the species belongs. Note how the following titles are more informative in the revised versions.

Vague	Mating Frequency in Butterflies
Vague	Mating Frequency in *Papilio*
Specific	Mating Frequency in Butterflies of the genus *Papilio*

The first title is too general, unless the writer is discussing *all* butterflies; the second will be confusing to readers unfamiliar with insect (and especially butterfly) taxonomy.

Vague	Effect of Hormones and Vitamin B on Gametophyte Development in a Moss
Specific	Effect of Hormones and Vitamin B on Gametophyte Development in the Moss *Pylaisiella selwyni*

The first title conveys the purpose of the study, but we are left wondering *which* moss is being studied.

Vague	Pollination, Predation, and Seed Set in *Linaria vulgaris*
Specific	Pollination, Predation, and Seed Set in *Linaria vulgaris* (Scrophulariaceae)

Even botanically inclined readers may not be familiar with this particular species. Including the family (Scrophulariaceae) after the scientific name provides the useful information that this plant is in the Snapdragon Family.

Vague	Paternity Assurance in the Mating Behavior of *Abedus herberti*
Specific	Paternity Assurance in the Mating Behavior of a Giant Water Bug, *Abedus herberti* (Heteroptera: Belostomatidae)

As in the preceding example, the first title of this pair is not informative enough for most readers. Is this organism a mammal? A fish? A bird? The revised version tells us that it is an insect belonging to the order Heteroptera and family Belostomatidae.

Note that there are some species or larger groups whose scientific names do *not* need to be specified in the title, although it is not incorrect to do so. These organisms include those whose common names are familiar, well established, and even standardized (such as American birds), and organisms that are well studied or familiar to most scientists, such as cats or rats. However, it is still necessary to include precise taxonomic information in the text of the paper.

Finally, there are some species — for example, *E. coli* and *Drosophila* — that are well known by their *scientific* names, which can appear without further explanation in the title.

■ Avoid specialized terminology, coined words, and most abbreviations.

As in a published paper, your title should be meaningful on its own; otherwise it will confuse or discourage potential readers. Write the title with your audience in mind. Use only those terms likely to be familiar to most readers — for example, your instructor and your classmates. Avoid esoteric or highly technical language, and do not use abbreviations except for widely recognized ones (such as DNA, RNA, or ATP).

ABSTRACT

■ Summarize the major points of the paper.

The abstract is a short passage (usually 250 words or less) that appears just after the title and author(s) and summarizes the major elements of the paper: objectives, methods, results, and conclusions. It is typically written as a single paragraph. In the case of published works, a good abstract helps researchers assess the relevance of a paper to their own research, and thus decide whether or not they should read it completely. Scanning the abstracts of papers is also one way that scientists, when pressed for time, keep abreast of recent literature.

Although readers see the abstract first, it is easiest to write it last, once you have a good overview of the paper. One way to write an abstract is to list, one by one, all the important points covered in each section of the paper. Write complete sentences if you can. Make successive revisions, paring the list down bit by bit, omitting peripheral topics and details, until you have revealed the "skeleton" of the study.

■ Be specific and concise.

As a summary, an abstract must be both informative and brief. Avoid general, descriptive statements that merely hint at your results or serve only as a rough table of contents. Consider every sentence, every word. Could you say the same thing more economically?

The following abstract from a published paper illustrates how the essential points of a study can be summarized in a single concise paragraph:

> The action of melanin-concentrating hormone (MCH) on melanophores was studied in 27 teleost species. MCH caused melanosome aggregation in all teleosts studied, including two siluroid catfish in which melanin-aggregating nerves are known to be cholinergic. In most fish, the minimal effective concentration of MCH was estimated to be 10 μM, while in three swellfish examined, it was higher than 10 μM. The mode of action of the peptide was identical in either adrenergically or cholinergically innervated melanophores. It may act through specific receptors on the melanophore membrane. These results suggest that MCH may be a biologically active hormone common to teleosts.
>
> (Nagal, Oshima, and Fujii 1986, p. 360)

The abstract below is too general: the student writer has left out important information and refers only vaguely to what was done, how, and why.

> This report is a study of the algae in Lebanon Reservoir, using a variety of collecting techniques. Over the study period, there was a gradual replacement of Cyanophyceae by Bacillariophyceae as a result of certain factors, such as temperature. Planktonic communities differed from attached communities because of morphological and physiological characteristics of the algal species in each community. There were marked differences in algal composition between the epilimnion and hypolimnion.

This abstract could be improved if the writer used more specific language. As it stands, we are left with unanswered questions. What kind of collecting techniques were used? Were there other important factors besides temperature? What "morphological and physiological characteristics" did the writer study? And so on. We also do not know where Lebanon Reservoir is, or why the author did the study in the first place.

The abstract below is unnecessarily wordy:

> This paper reports studies on the territorial behavior patterns shown by males of the dragonfly species Plathemis lydia at a small pond in the vicinity of Earlville, New York. A total of 51 male dragonflies were marked with small spots of enamel paint applied

> to the abdominal portion of the body and were observed under natural conditions during the month of June 1987. It was found that males showed a strong tendency to defend individual areas at the pond. These males chased away other males of their species, and also, on occasion, even males of other species of dragonflies. Threat behavior, as opposed to behavior involving physical contact, was the most common aggressive behavior displayed by males defending their areas against intruders. Males typically remained at the same area of the pond for two hours or longer on a single day. Also, they usually were observed returning to the same area on successive days. These observations suggest the possibility that male territorial behavior in this species serves the adaptive function of enabling the owners to monopolize particular areas. These areas may be visited by females who are in the act of seeking mates.

This abstract used 192 words to describe what the same abstract, written more concisely, can express in 104 words:

> I studied the territorial behavior of male dragonflies (Plathemis lydia) at a small pond near Earlville, New York. Fifty-one males were marked with enamel paint on the abdomen and observed under natural conditions during June 1987. Males defended individual areas from male conspecifics, and occasionally from males of other species. Aggressive interactions generally involved threat behavior rather than physical contact. Territory owners typically remained at the same area for at least two hours and returned to the same location on successive days. Territorial behavior in this species may be adaptive for males, enabling them to monopolize areas likely to be visited by females seeking mates.

In the revised version, the author has replaced long, wordy constructions with shorter, more economical wording: for example, "the territorial behavior patterns shown by males of the dragonfly species *Plathemis lydia*" has become "the territorial behavior of male dragonflies (*Plathemis lydia*)." Many unnecessary words ("a total of," "it was found that," "these observations suggest the possibility that") have been omitted. The resulting Abstract conveys the same information much more effectively. (See pp. 108–112 for a more complete discussion of wordiness.)

The Abstract below is too long because the author included more information than was necessary. What parts should be left out?

> The algal succession occurring with autumn overturn was studied in the phytoplankton community of a hard water pond near Madison, New Jersey. The pond can be sampled easily from all sides and offers excellent opportunities for ecological research. This study was done as a field project for Biology 314. The dominant species in October 1986 were *Dinobryon sertularia*, *Scenedesmus quadricauda*, *Cryptomonas erosa*, and *Cyclotella michiganiana*. The first three species showed significant declines in December, after fall overturn, whereas the last species became much more prominent.

We do not need to know that the pond is a good place to do research, or that the study was done for a particular biology course.

■ Make the abstract able to stand alone and still make sense to the reader.

Like the title, the Abstract of a published work will be read by many people who may not read the text of the paper. These include researchers using references such as *Biological Abstracts* to locate relevant references on a topic. Therefore, the Abstract must be independent: readers should be able to understand it without being familiar with the details of the study. Omit abbreviations (except for widely known ones such as DNA), and use only those technical terms likely to be familiar to your audience. If certain abbreviations or specialized terminology are absolutely necessary, then explain them. Do not refer to materials that may be inaccessible to your readers, such as your tables and figures, and avoid references to other literature, if possible.

The following Abstract, taken from a student research report, is difficult to understand without having first read the paper:

> This paper describes a behavioral and computer simulation study of cognitive differences between five-year-old boys and girls presented with tasks of either Group I, II, III, or IV difficulty level. We followed a modified version of the method developed by Wolf (1955, 1977). When compared to boys, girls scored higher on what we designated "prepared tests," whereas boys scored higher on "stimulus-response tests" (Fig. 3). We concluded that. . . .

The author of this abstract provides little information we can understand without having first read the paper. We have no way of knowing what "Group I, II, III, or IV" stands for, what Wolf's method was, what the author's definitions of "prepared" and "stimulus-response tests" were, or what Figure 3 is all about.

Now look again at the abstract by Nagal, Oshima, and Fujii (1986) (p. 10). Notice that the abbreviation MCH is explained and the language is straightforward. The few technical terms that are used (for example, "melanophores" and "cholinergic") are likely to be familiar to most readers in this field.

INTRODUCTION

The Introduction of a research paper sets the stage for your scientific argument. It places your work in a broad theoretical context and gives readers enough information to appreciate your objectives. A good Introduction "hooks" its readers, interesting them in the study and its potential significance. Thus you, as a writer, must have a firm grasp of the aims, principal findings, and relevance of your research. You may find that the Introduction is easiest to write after you have drafted the Materials and Methods, Results, and Discussion sections and have a clearer understanding of just what you are introducing.

■ Orient the reader by summarizing pertinent literature in your field.

An effective way to organize the Introduction is to proceed from the general to the specific, starting with a review of current knowledge about the topic and narrowing down to your specific research problem. Introduce key concepts, define specialized terms, and explain important hypotheses or controversies. As you do so, document your writing by citing key references in the field — more general ones first, followed by studies closest to your own research. In this way you can sketch out a framework for the study, orient your readers, and prepare them for what is to follow.

Do not make the Introduction *too* broad or too detailed. This is not the place to show off your knowledge about the subject, list every available reference, or repeat material found in any elementary text. Most published papers have short introductions (often only a few paragraphs) because the writer is addressing readers with backgrounds similar to his or her own. It wastes journal space and the reader's time to give an exhaustive literature review. Similarly, in a paper for a course, write for your classmates and your instructor — people with at least a beginning knowledge of the subject. Discuss only the most relevant concepts and references, and get quickly to the point of the paper.

■ Explain the rationale for the study and your major objectives.

After explaining the broad theoretical context, you are ready to clarify the special contribution your own study makes. How does your work fit in with that of other researchers? What special problem does your study address? What *new* information have you tried to acquire? Why? How? In other words, why (apart from course requirements) are you writing the paper in the first place? Most authors end the introduction by stating the *purpose* of the study. For example:

> The purpose of this study was to describe the dominant fungi associated with decomposing leaf litter in a small woodland stream.
>
> In this paper, I report laboratory observations on the effects of crowding on the behavior of juvenile Atlantic salmon (*Salmo salar*).

Remember to clarify the purpose of your paper by first explaining the rationale for the study. What needs to be improved in the following Introduction?

```
The purpose of this project was to study the algae
at two different sites in Payne Creek, near Still-
son, Florida.  Chemical and physical parameters were
also considered.  The study involved algal distribu-
tion in several microhabitats of both stream and
marsh environments.  This project was part of a
larger class study for Biology 341.
```

A major problem here is that the writer has not linked the study to a broader conceptual framework. She begins abruptly with a vague statement of her objectives, with no explanation of why these are important. The passage also contains irrelevant information: it is not necessary to state that the study was part of the requirements for a particular course.

By contrast, the writer of this Introduction gives the reader a clear idea of what the objectives are and why they are important:

> It is well known that males of many species of dragonflies (order Odonata) guard their mates after copulation while the eggs are being laid. In some species the male hovers over the female and chases away other males; in others, the male remains physically attached to the female as she moves around the breeding site. Recently, many authors have discussed the adaptive significance of mate guarding, particularly with respect to ways in which it may increase the reproductive success of the guarder (Alcock 1979, 1982; Sherman 1983; Waage 1979, 1984). A more complete understanding of the evolution of this behavior is dependent on detailed studies of many odonate species. In this paper, I describe mate guarding in Sympetrum rubicundulum and discuss ways in which guarding may be adaptive to males.

Some authors end the Introduction by summarizing their major *results* in a sentence or two. This tactic gives readers a preview of the major findings and may better prepare them for the scientific argument that follows. Other writers, along with some journal editors, criticize this practice, arguing that results are already covered in their own section and in the Discussion and Abstract. Ask your instructor what he or she prefers.

MATERIALS AND METHODS

■ Include enough information so that your study could be repeated.

Your methodology provides the context for evaluating the data. How you made your measurements, what controls you used, what variables you did and did not consider — all these things are important in molding your interpretation of the results. The credibility of your scientific argument depends, in part, on how clearly and precisely you have outlined and justified your procedures.

Furthermore, one of the strengths of the scientific method is that results should be reproducible using similar materials and methods. It is not unheard of for a scientist to repeat someone else's experiment and get different results. These conflicting data then point to factors that may have been overlooked, perhaps suggesting different interpretations of the data.

Finally, a complete and detailed Methods section can be enormously helpful to others working in the same field who may need to use similar procedures to address their own scientific problems.

What kinds of information should you include? See the guidelines below.

Materials

1. Give complete taxonomic information about the organisms you used: genus and species names as well as subspecies, strains, and so on, if necessary. Specify how the organisms were obtained, and include other information pertinent to the study, such as age, sex, size, physiological state, or rearing conditions.

> *Cladosporium fulvum* race 4 was obtained from Dr. George Watson, Department of Biology, Colgate University, Hamilton, New York. Stocks were maintained in sterile soil at 4°C and were increased on V8 juice agar at 25°C in the dark. . . .

> Adult American chameleons (*Anolis carolinensis*), purchased from a local supplier, were used for all experiments. They were kept in individual terraria (30 cm × 30 cm × 30 cm) for at least seven days prior to the start of any study. They were provided with a constant supply of water and were fed crickets, mealworms, and other insects every two days. All chameleons were exposed to 15 h of fluorescent light daily (0800–2300 h), and air temperatures were kept at 30°C. . . .

2. If you used human subjects, give their age, sex, or other pertinent characteristics. Biologists submitting papers for publication may need to demonstrate that subjects have consented to be involved in the study.

3. Describe your apparatus, tools, sampling devices, growth chambers, animal cages, or other equipment. Avoid brand names, unless necessary. If some materials are hard to obtain, specify where you purchased them.

4. Specify the composition, source, and quantities of chemical substances, growth media, test solutions, and so on. Because they are more widely understood, use generic rather than brand names.

> Sodium citrate, sodium pyruvate, and hydroxylamine were obtained

from Sigma Chemical Company, St. Louis, Missouri. All chemicals were of reagent grade. . . .

For the first series of electron microscopy studies, tissue samples were fixed in 2.8% ultrapure glutaraldehyde, 0.57 *M* glucose and 0.10 *M* sodium cacodylate. . . .

5. If detailed information about any of the materials is available in a standard journal, then avoid repetition by referring the reader to this source.

I used an intermittent water delivery system similar to that described by Lewison (1950). . . .

Tryptone-yeast extract broth (Pfau 1960) was used to cultivate bacterial strains. . . .

Methods

1. Describe the procedures in detail. Do not forget crucial details such as temperature conditions, pH, photoperiod, duration of observation periods, sampling dates, and arbitrary criteria used to make particular assessments or measurements. If you used a method that has already been described in a standard journal, you need not repeat all this information in your own paper; just cite the reference.

Mycelia were prepared using the fixation and embedding procedures described by Khandjian and Turner (1971). . . .

However, if the reference is hard to obtain (for example, *The Barnes County Science Newsletter*), or if you altered someone else's methods, then supply full information about your procedures.

2. For field studies, specify where and when the work was carried out. Describe features of the study site relevant to your research, and include maps, drawings, or photographs where necessary. If published information already exists on the area, cite sources.

This study was conducted during June and July 1986 at Bog Pond, 3 km northwest of Barrow, West Virginia. The general habitat of this pond has been described elsewhere (Needham 1967, Scott 1981). The pond is permanent and contains floating and emergent vegetation (mostly sedges, rushes, and algae). It has an area of approximately 1.5 ha. . . .

3. Commonly used statistical methods generally need no explanation or citation; just state for what purpose you used them. If you used less familiar or more involved procedures, cite references explaining them in detail and give enough information to make your data meaningful to the reader.

■ Organize your material logically.

Because the Materials and Methods section contains so many important details, it is easy to forget some of them, particularly because you are so familiar with the subject. It is also easy to let this section become a confusing, rambling conglomeration of details, with little unity or coherence. Organize this section carefully using an outline, a list, or a plan along with the detailed notes you compiled while doing the research.

A typical approach is to begin with a description of important materials (study species, cell cultures, and so on) and to move on to the methods used to collect and analyze the data. Field studies often start with a description of the study site. Describe your procedures in a logical order, one that corresponds as closely as possible to the order in which you discuss your results. You may also group related methods together.

Remember that the Materials and Methods section is still part of your text and must be readable. Do not let your paragraphs become disorganized collections of choppy sentences, as in the example below:

FAULTY Golden hamsters (*Mesocricetus auratus*) used for this research were adult males. Temperature conditions were kept at 22–24°C. Animals were fed Purina chow. Hormonal studies were performed on 23 individuals. The photoperiod was 16 h. Animals were housed with littermates of the same sex, and feeding was once each day. All hamsters had been weaned at three weeks.

REVISED Hormonal studies were performed on 23 adult male golden hamsters (*Mesocricetus auratus*). All had been weaned at three weeks and housed with littermates of the same sex. They were reared under conditions of 22–24°C and a photoperiod of 16 h, and fed Purina chow once daily.

The revised passage is easier to read and understand. Short, choppy sentences have been combined, and related points have been pulled together. The same information is conveyed, but using fewer words and a more organized style.

If the Materials and Methods section is longer than several paragraphs and involves lengthy descriptions of several topics, you may wish to use *subheadings* (perhaps taken directly from your outline) that break the text into clearly labeled sections (for example, Study Area, Sampling Methods, and Data Analysis; or Test Water and Fish, Testing Conditions, Chemical Analyses, and Statistical Analyses; or Plant Material, Morphometry, Light Microscopy, and Electron Microscopy). Make your subheadings general or more specific, depending on the type and amount of information you need to relate. If you use many specialized or invented terms to report your results, these may be put in the Methods section under Definitions. Using subheadings makes your text easier to write and to read, and it prods your memory for stray details on all aspects of the study.

■ Use specific, informative language.

Give your readers as much information as you can. Replace vague, imprecise words with more specific ones, and quantify your statements wherever possible.

VAGUE	I observed some monkeys in a large outdoor enclosure and others in small, individual indoor cages.
SPECIFIC	I observed 13 monkeys in an outdoor enclosure (10 m × 8 m × 12 m) and 12 others in individual indoor cages (1 m × 2 m × 1 m).
VAGUE	Several pits were dug at each forest site, and soil samples were collected from three different depths in each pit.
SPECIFIC	Four randomly located pits were dug at each forest site, and soil samples were collected from three depths at each pit: 0–5 cm, 6–11 cm, and 12–17 cm.
VAGUE	Root nodule tissue was stained with a number of histochemical reagents.
SPECIFIC	Root nodule tissue was stained with toluidine blue, Schiff's reagent, and aceto-orcein.
VAGUE	Every nest was checked frequently for signs of predation.
SPECIFIC	Every nest was checked twice daily (at 0800 and 1600 h) for signs of predation on eggs or nestlings.

■ Omit unnecessary information.

Include only those procedures directly pertaining to the results you plan to present. Do not get carried away in your desire to include all possible details. The reader is not interested in superfluous details or asides, and in published articles such material justs wastes space and raises printing costs. The following example includes many unnecessary details:

> Fathead minnows were collected from Lost Lake, near Holmes, North Carolina, and transported back to the third-floor laboratory in large white pails. On the following Tuesday morning, skin for

> histological examination was taken from the dorsal part of the fish just behind the anterior dorsal fin.

Do we really need to know that the fish rode home from the lake in large white pails? Or that they were taken to that particular laboratory? Or that histological studies were done on a Tuesday?

Notice how much clearer the following example becomes when superfluous details are removed:

FAULTY After considering a variety of techniques for determining the sugar content of nectar, I decided to use the method developed by Johnson (1955), because it seemed straightforward and easy to follow, especially for someone with a poor mathematical background.

REVISED I used Johnson's (1955) method to measure the sugar content of nectar.

Do not confuse readers of your Materials and Methods section by insisting they jump ahead to your Results.

> To quantify the fright response, I observed 10 groups of fish, each composed of five individuals, and recorded the number of movements per 5-minute interval (Fig. 1).

Remember that you are dealing strictly with procedures here. When readers encounter Figure 1 later in the paper, they will know enough to consult the Methods section if they need to.

RESULTS

■ Summarize and illustrate your findings.

The Results section should (1) *summarize* the data, emphasizing important patterns or trends, and (2) *illustrate* and *support* your generalizations with explanatory details, statistics, examples of representative (or atypical) cases, and references to tables and/or figures. To convey the results clearly, your writing must be well organized. Present the data in a logical order, if possible in the order in which you described your methods. Use the *past* tense (see pp. 116–117). If the Results section is long and includes many different topics, consider using subheadings to make the text easier for the reader to grasp.

The following paragraph is from a student paper on reproductive behavior in a species of damselfly. Notice how the author begins with a general statement and then supports it with quantitative data, a graph accompanying the text, and a selected example.

Observations of 21 marked males showed that the number of matings per day varied among individuals. The number of matings per male ranged from 0–10 per day, with a mean of 6.5 (SD = 3.2). Males occupying territories with abundant emergent vegetation encountered more females and mated more frequently than males occupying sparsely vegetated areas (Fig. 3). One male, whose territory consisted entirely of open water, obtained no matings during the five days he was observed there.

■ Do not interpret the data or draw major conclusions.

The Results section should be a straightforward *report* of the data. Do not compare your findings with those of other researchers, and do not discuss why your results were or were not consistent with your predictions. Avoid speculating about the causes of particular findings or about their significance. Save such comments for the Discussion.

The following passage is from the Results section of a student paper on the algae present in a small stream. The writer begins with a statement of her findings (the first sentence), but then continues with interpretive material that really belongs in the Discussion.

The epilithic community was dominated by Achnanthes minutissima (Table 4). The abundance of this species at the study site may be related to its known tendency to occur in waters that are alkaline (Patrick 1977) and well oxygenated (DeSeve and Goldstein 1981). Its occurrence may also have been related to. . . .

■ Integrate quantitative data with the text.

If the Results section includes tables or figures (discussed fully in Chapter 3), be sure to refer to *each* of these in the text. Do not excessively repeat in the text what is already shown in a table or figure. On the other hand, don't restrict yourself to passing comments ("Results are shown in Table 1"). Instead, point out the most important information or patterns and discuss these in the context of related data.

> Reproductive activity was closely related to time of day. Both males and females began to arrive at the breeding site in late morning, and the density of both sexes was highest between 1300–1500 h (Fig. 3). Data on air temperatures (Fig. 4) suggest that. . . .

Do not automatically assume that anything involving numbers *must* be tabled or graphed in order for it to look important or "scientific." Unnecessary tables or figures take up space and waste the reader's time. Once you are sure which results are important and why, you may find that many can be summarized easily in a sentence or two. Following are some general guidelines about how to present numerical results verbally.

1. An average or mean value is often accompanied by the standard deviation, which gives the reader a sense of the variability of the data within the sample. Where consideration of the highest and lowest values is important, the range may be reported along with the sample size (number of observations). If you wish to show how reliable a sample mean is as an estimator of the population mean, give the standard error. See page 134 for useful references on the computation of these and other basic statistics.

The following sentences illustrate how statistical results can be integrated smoothly with the text:

> The 15 caterpillars in Group 3 averaged 2.1 cm in total length (range = 1.6 − 2.6).

> In the bullfrog choruses at Werner Lake, the mean size of central males was 140.42 mm (SD = 7.45, N = 12).

> You may also denote the standard deviation as follows:

> 140.42 ± 7.45 mm.

2. When reporting the results of commonly used statistical analyses, you need not describe the tests in detail nor give all your calculations. Generally, you need to report only the major test statistics, along with the significance or probability level (see pp. 23–24).

> Analysis of variance showed significant variation among females in mean egg mass ($F_{[29.174]} = 25.4$, $P < .001$).

3. As shown above, conventional abbreviations and symbols are used to report data succinctly within sentences. Some common ones are listed on the inside back cover of this book.

4. When do you write out the words for numbers, and when do you use numerals? The general rule is to use numerals when you (1) report statistics; (2) give quantitative data using standard units of measurement; and (3) refer to dates, times, pages, figures, and tables.

> The experiments were conducted between 1200 and 1500 h on September 8, 9, and 10, 1987.
>
> Overlap of burrow groups ranged from 0 to 35%, and averaged 12% (Fig. 2 and Table 4).

In other situations involving numbers, use words for values of one through nine, and numerals for larger values. If you have a series that includes both values, use numerals for everything.

> Solitary individuals were two to three times more likely to encounter predators than were members of any of the five colonies.
>
> We followed the movements of 3 marked individuals, 22 marked pairs, and 14 unmarked pairs.

Do not *begin* a sentence with a number. Either write out the number or revise the sentence. If the number is part of a chemical term, it cannot be spelled out and the sentence should be reworded.

FAULTY	12 out of 425 eggs were cracked at the start of the first experiment.
REVISED	Twelve out of 425 eggs. . . .
FAULTY	6-mercaptopurine was used to inhibit mitosis.
REWORDED	Mitosis was inhibited by 6-mercaptopurine.
REWORDED	I used 6-mercaptopurine to inhibit mitosis.

5. Watch your wording when you report quantitative results. Several commonly used words have restricted statistical meanings when they appear in scientific writing; do not use them loosely.

Significant: In popular usage, this word means "meaningful" or "important." Scientists use this term in a more restricted sense to refer to *statistically significant* (or nonsignificant) results, after having conducted appropriate statistical tests:

> Observed frequencies of turtles obtaining food differed *significantly* from expected frequencies ($\chi^2 = 58.19$, df $= 8$, $P < .001$).

Because the above results were significant, we know that the difference between observed and expected frequencies is probably a valid one. In this case, the likelihood of a difference of at least this magnitude occurring by chance alone is $< .001$, or less than one in a thousand. Note that the leading zero is often implied but not typed in probability values. Thus < 0.001 is given as $< .001$.

When reporting the results of statistical tests, you always need to specify the significance level, as indicated by a probability value (P) such as the one above. Results associated with probabilities greater than .05 are generally considered *not* significant.

Correlated: In popular usage, two entities that are correlated are related to each other in some way. In scientific writing, the term *correlated* is used in conjunction with certain statistical tests (correlation analyses) that provide a measure of the strength of relationship between two variables.

> Female size was not significantly *correlated* with the percentage of abnormally developing embryos in an egg mass ($r = 0.29$, $P <$ .05, $N = 23$).

Note that correlation does *not* allow you to automatically assume a cause-and-effect relationship between the variables. It merely describes the extent to which they co-vary. For example, if variables X and Y are correlated, high values of X tend to be associated either with high values of Y (positive correlation) or with low values of Y (negative correlation).

Random: In popular usage, this word means "haphazard" or "with no set pattern." However, scientists use this term to refer to a *particular* statistically defined pattern of heterogeneous values. If you write, "Subjects were assigned *randomly* to either Group A or Group B," then you must have used a table of random numbers or some other accepted method to make your group assignments truly random.

■ Omit peripheral information and unnecessary details.

Most scientists amass far more data than they ever present to their readers. Similarly, even in a week-long project, you may have more results than you know what to do with. You may, understandably, feel reluctant to part with any of them. However, if you pack every last bit of information into the paper, you may lose sight of why you did the study in the first place. You must learn to present results selectively, to choose the data that are relevant to your hypothesis because they are either consistent or inconsistent with it. The data you present as results are the same data you will use to support your conclusions. You do not want to confuse the reader (or yourself) by including irrelevant information.

For example, the passage below is from a student field project on the diversity of vascular plants at a small pond.

> As shown in Figure 1, the shoreline of Hicks Pond was generally predominated by grasses and sedges. Cat-tails occupied a small area on the northern end, and goldenrods (Solidago spp.) were present in scattered groups along the eastern shore. A large popu-

> lation of red-spotted newts (Motopthalmus viridescens) was also spotted in this area, along with several species of ducks. Species diversity was highest at Site D (see Table 1), which was drier than most areas and contained some of the same plants as in the surrounding hayfield.

The third sentence in this paragraph should be omitted, because such zoological observations, however interesting, are irrelevant in this context.

Similarly, do not clutter the Results (or any other) section of the paper with irrelevant general statements of aims or intent. The following sentence, for example, should have been omitted from the student paper in which it appeared:

> To present the results of this study, I will first examine all of the relevant physiological factors and then discuss the findings of each feeding experiment.

You need not explain how you will proceed. If your writing is coherent and well organized, readers will follow your train of thought.

DISCUSSION

■ Interpret your results, supporting your conclusions with evidence.

In the Results section you reported your findings; now, in the Discussion, you need to tell the reader what you think your findings mean. Do the data support your original hypothesis? Why or why not? Refer to your data, citing tables or figures where necessary; use these materials as evidence to support your major argument or thesis. Here is the place, too, to discuss the work of other researchers. Are your findings consistent with theirs? How do your results fit into the bigger picture?

■ Do not present every conceivable explanation.

Sometimes beginners feel obliged to think of every possible way to interpret their results. The reader, swamped by explanations (most of them highly speculative), will quickly lose faith in the author. Look at this excerpt from a student Discussion section:

> The dramatic decrease in <u>Ochromonas</u> in December may have been related to the formation of cysts (perhaps overlooked in the water samples taken at that time). It is also possible that this alga died off suddenly because of some environmental stress, such as a prolonged period of cold temperatures or a sudden chemical change in the water. Light intensity and photoperiod are other potential factors. Finally, the December samples were not analyzed until the day after they had been collected because two members of our lab group were sick, so by this time the algal populations may have been different from the ones at the field site.

Remember that your task in the Discussion is to argue on behalf of the most plausible interpretations, based on the evidence available to you. To do so, you must be selective, focusing on explanations that have the greatest bearing on your study. Omit wild guesses and irrelevant asides.

■ Recognize the importance of "negative" results.

Experiments do not always have to confirm the presence of major differences, a strong relationship between two variables, or a conspicuous trend or pattern to be considered good science. Sometimes you may find that there is no significant difference between two groups, or no relationship between a particular process and, say, temperature or some other factor you have been investigating. Such "negative" results still constitute respectable science, and they still need explanation. Thus, be alert to unexpected findings, and don't automatically conclude that the experiment is a failure or that you've made a mistake. Cases that do not conform to the expected pattern might represent something important — perhaps a new or altered focus for your study.

■ Make your prose convey confidence and authority.

Show that you are knowledgeable about your subject and take responsibility for your conclusions. Do not hedge or apologize. The tone in the following student examples is unnecessarily tentative:

```
This is just a preliminary study of the social be-
havior of the guppy (Poecilia reticulata), but it may
possibly shed some light on the subject and serve as
a base for more conclusive work in the future.
```

Would you want to read this paper? Perhaps not, if it is as preliminary and inconclusive as the author implies.

```
These results seem to suggest the possibility of at
least some connection between ferric chloride and
increased disease resistance.
```

This writer seems afraid to take a stand. What *do* the data show? If they suggest a connection between ferric chloride and disease resistance, then the writer should say so: "These results suggest a connection between ferric chloride and disease resistance."

```
These findings may not be very accurate because of
my limited experience, although they appear to be
consistent with the observations of Wilbur (1972)
and Morris (1978).
```

The last writer, like many other students, obviously feels intimidated by the work of "experts." However, "beginners" often do very good scientific work. Do not weaken your conclusions by unnecessary references to your status as a novice. If you are reluctant to make a strong statement, ask yourself why. Perhaps your anxiety stems from reservations about the data or the way the experiment was conducted. Perhaps there are certain variables you have not considered or certain assumptions that don't ring true. These factors *are* worth considering, and it is important to sort them out from any general insecurities you have about yourself as a scientist or a writer.

■ Use a coherent, logical organization.

Instead of proceeding from the *general* to the *specific*, as the Introduction does, the Discussion moves from the *specific* to the *general*. There is no one right way to put together a Discussion, but the following plan is a common and effective one.

1. Start by drawing attention to your major findings without excessively repeating the Results. The reader has just looked at your data, so you do not have to describe them all over again. Beginning writers often start the Discussion by dredging up material from the Introduction; this approach, too, just adds redundancy to the text.

Focus the reader's attention on the most important findings, patterns, or trends. For example, here is the beginning of the Discussion section from a paper by Young, Greenwood, and Powell (1986, p. 400):

> Although interactions between molluscs and their potential predators have been studied extensively (for reviews see Thompson 1960, Edmunds 1966, Todd 1981, Faulkner and Ghiselin 1983), these data are the first in which predator-prey interactions have been documented for a large number of potential predators of intertidal Onchidiids. In addition, our data indicate that the defensive secretion of *Onchidella borealis* has an effect on the distribution of the predatory seastar *Leptasterias hexactis*. The results indicate that *O. borealis* does not fire its repugnatorial glands in response to all potential predators, nor do all potential predators demonstrate flight behaviors in response to the glandular secretions of *O. borealis*.

2. Ask yourself what causes may underlie the major trends or phenomena you have described in the paper. If there are conflicting, problematic, or unexpected results, suggest explanations. Here is an excerpt from a student research report:

> The unusual presence of Chaenorrhinum minus in an untended garden (Site 5), instead of in its usual habitat of railroad cinder ballast, may be related to the use of railroad ties as decorative borders around the garden. Seeds from C. minus may have been lodged in crevices in the railroad ties, transported to the garden, and dislodged as the ties were being set in place. However, it is still unusual that C. minus was growing so well in a moist, shaded area among many other plants. My findings suggest that C. minus generally is a relatively poor competitor and grows best on dry, exposed, gravelly sites where few other plants are found.

3. Compare your findings with the work of other researchers. Are your results similar to theirs? At this point you can begin to supplement your own evidence with relevant findings from other studies, showing the reader how your work is part of a broader framework. Be sure to look at the subject fairly and honestly. If some authors obtained results different from yours, point this out and suggest explanations for the differences. The following example is from McMillan and Smith (1974, p. 52).

> The activities of males after spawning interested other observers of fathead minnows. Miller (1962) reported that a parental male

> positioned himself in the mouth of the cavity below the eggs and constantly fanned with his pectoral fins. This behavior has not been noted by other workers, nor has it been observed during the present study. It is possible that Miller was watching males hovering (our definition) below their eggs. In that case it is doubtful that the weak pectoral fin movements employed in hovering could substantially aerate the eggs — the function Miller attributed to "fanning."

4. End with more far-reaching predictions, interpretations, and conclusions. Can you generalize from your specific findings to other situations? How does your work contribute to an understanding of the broader topic? If you can end with a firm statement, as the student example below does, you give the reader a satisfying sense of closure.

> In conclusion, cannibalism of eggs by larvae of the butterfly Euphydryas phaeton occurred commonly under natural conditions, even within relatively small colonies. These results are in agreement with Wilson's (1975) suggestion that detailed studies of cannibalism in animals may show it to be more common than is usually supposed. The finding that cannibalistic larvae grew more rapidly than noncannibalistic ones suggests that cannibalism may be an important factor in larval development, especially when food supplies are scarce. This nutritional benefit of cannibalism may have long-reaching effects. If successful laboratory culture methods can be developed for E. phaeton, we can further explore the relationship between cannibalism and individual fitness.

ACKNOWLEDGMENTS

In published biological papers, a short Acknowledgments section usually comes between the Discussion and the Literature Cited sections (Chapter 4). Here you express thanks to people who assisted you with the research itself or with the preparation of the paper. Scientific ethics dictate that you first obtain the permission of all those you intend to acknowledge, preferably by showing them a copy of that section of the paper. This is because it is conceivable (though probably unlikely) that some people who offered you help may not wish to be associated in print with the paper, or because they may object to the wording you used to refer to them. See page 35 for an example of an Acknowledgment.

SAMPLE RESEARCH PAPER

The following paper reports the results of a student's independent research project for a mycology course.

The Effect of Sucrose Concentration
on the Formation of Appressoria
in Colletotrichum lindemuthianum
(Coelomycetes, Melanconiales)

Andrea Rider

Biology 214
May 5, 1988

Include title, author, course, and date on title page for student paper. Include address of author's institution on manuscripts submitted for publication. Do not number title page, but consider it page 1.

Underline (italicize) genus and species. Capitalize genus only.

Summarize paper's most important contents using past tense. Use present to suggest a general conclusion.

Label and begin on same page as Abstract. In longer paper or journal manuscript, start new page.

Spell out *Colletotrichum* once, then abbreviate it when preceding species name.

For background, briefly review pertinent literature. Name-and-year method is used here to cite sources. Use present tense to refer to established knowledge.

ABSTRACT

Spores of the fungus Colletotrichum lindemuthianum were germinated under different sucrose concentrations in the range 1 x 10^{-1} to 1 x 10^{-7} M. The formation of germ tubes with appressoria, as opposed to germ tubes only, decreased with increasing sucrose concentration. This finding may be significant in light of the leaching of sugars onto plant surfaces. Export of such substances may decrease the frequency of appressoria formation and reduce the likelihood of infection by parasitic fungi.

INTRODUCTION

The fungus Colletotrichum lindemuthianum causes anthracnose, a disease of green beans (Phaseolus vulgaris). Once on the surface of a bean seedling or pod, a germinating spore of C. lindemuthianum typically forms a short germ tube that swells to form a spherical structure called an appressorium. The fungus then penetrates the host tissues from the appressorium. The morphology and roles of appressoria have been reviewed by Emmett and Parberry (1975). In laboratory experiments using etiolated hypocotyls of P. vulgaris sprayed with aqueous suspensions of spores of C. lindemuthianum, appressoria develop from most spores, usually on the ends of very short germ tubes (Skipp and Deverall 1974). Longer germ tubes (ones longer than about four times the average length of a spore) do not ever form appressoria and thus fail to penetrate the host tissue (Arnold and Lyons 1982).

In culture on agar media supplemented with sugars as carbon source, C. lindemuthianum spores produce germ tubes but no appressoria. However, spores suspended in distilled water do form appressoria (R. M. Arnold, pers. comm.) These observations suggested a need to determine the effect of nutrients in the culture medium on germination in this species. Arnold and Lyons (1982) showed that fewer appressoria are formed when spores of C. lindemuthianum are incubated in sucrose solutions. In this paper, I describe a more detailed study on the germination of this fungus under increasing concentrations of sucrose.

With authors' permission, cite personal communications (any unpublished data). Except for works in press, omit these from Literature Cited.

Rationale for study, followed by paper's objectives.

MATERIALS AND METHODS

C. lindemuthianum was grown for 10 days on agar containing 35% bean juice. I made spore suspensions by flooding the petri plates with distilled water and gently rubbing the surface of the agar with a glass rod. The spores were washed by low-speed centrifugation in distilled water three times. Then they were mixed with serial dilutions of a stock sucrose solution (2×10^{-1} M) to give final sucrose concentrations in the range 1×10^{-1} to 1×10^{-7} M. I also prepared a control containing distilled water. A syringe was used to place 25-μl

Use active voice when appropriate; use passive voice to focus attention on materials, not yourself.

Appressoria in C. lindemuthianum 4

Describe procedures in enough detail for others to repeat study.

droplets of the spore suspensions in water or sucrose solutions on clean glass microscope slides. A cover glass was added to each and the slides were incubated in high humidity conditions for 24 h. After incubation, the preparations were examined microscopically. Spores in randomly selected fields of view were classified according to germination morphology (appressorium or germ tube only.)

RESULTS

Number figure even though only one is given. Identify important findings of graph.

As shown in Figure 1, the proportion of spores that formed appressoria at the ends of germ tubes decreased with increasing sucrose concentration. Sucrose concentration did not significantly affect the proportion of spores that germinated in a treatment ($\chi^2 = 9.26$, df = 7, $P > .05$).

DISCUSSION

Interpret results or compare with other studies. Summarize results of statistical test. Use *significant* only for statistical significance.

Briefly review pertinent findings. Do not excessively repeat Results.

Relate findings to others'. Shift to present tense to discuss previously published information.

In this study, the mode of germination in C. lindemuthianum was affected by nutrient characteristics of the culture medium, with the production of germ tubes without appressoria occurring at high sucrose concentrations. This finding is in accord with data of Arnold and Lyons (1982). Appressoria formation is characteristic of the parasitic phase of the fungus, while germ tube formation without appressoria is associated with a saprobic or nonparasitic phase (Leach 1923).

In nature, sugars and other organic molecules leach onto plant surfaces (Godfrey 1976), presumably through stomata or other

Appressoria in C. lindemuthianum 5

natural openings, in an apparently wasteful process. The findings reported in this paper suggest that, although these substances may encourage the growth of saprobic fungi, they may also make a potentially parasitic fungus grow only in a saprobic manner. Thus, the export of organic molecules to plant surfaces may have an adaptive value in helping to keep parasitic fungi outside plants.

End with broader significance of study.

ACKNOWLEDGMENTS

I thank Dr. R. M. Arnold for his advice during all stages of the project, and Ms. Jennifer Arnold for critically reviewing the manuscript.

With their permission, acknowledge those who assisted with project.

LITERATURE CITED

Include all references cited. Alphabetize sources by first author's last name.

Arnold, R. M., and B. M. Lyons. 1982. Laboratory investigations using the anthracnose disease of beans. Am. Biol. Teach. 44:51-55.

Emmett, R. W., and D. G. Parberry. 1975. Appressoria. Ann. Rev. Phytopathol. 13:147-167.

Godfrey, B. E. S. 1976. Leachates from aerial parts of plants and their relation to plant surface microbial populations. Pages 433-439 in C. H. Dickinson and T. F. Preece, eds. Microbiology of aerial plant surfaces. Academic Press, London.

Leach, J. G. 1923. The parasitism of Colletotrichum lindemuthianum. Minn. Agr. Exp. Sta. Tech. Bull. 14:1-44.

In article from edited collection, list page numbers followed by editor, title of book, publisher, place of publication.

Use initials for authors' first and middle names (with first author's name in reverse order) followed by publication date, title of article with only first word capitalized. Use conventional abbreviations for journal title. Give volume number followed by page numbers.

Skipp, R. A., and B. J. Deverall. 1974. Relationships between fungal growth and host changes visible by light microscopy during infection of bean hypocotyls (Phaseolus vulgaris) susceptible and resistant to physiological races of Colletotrichum lindemuthianum. Physiol. Plant Pathol. 2:357-374.

Appressoria in C. lindemuthianum 7

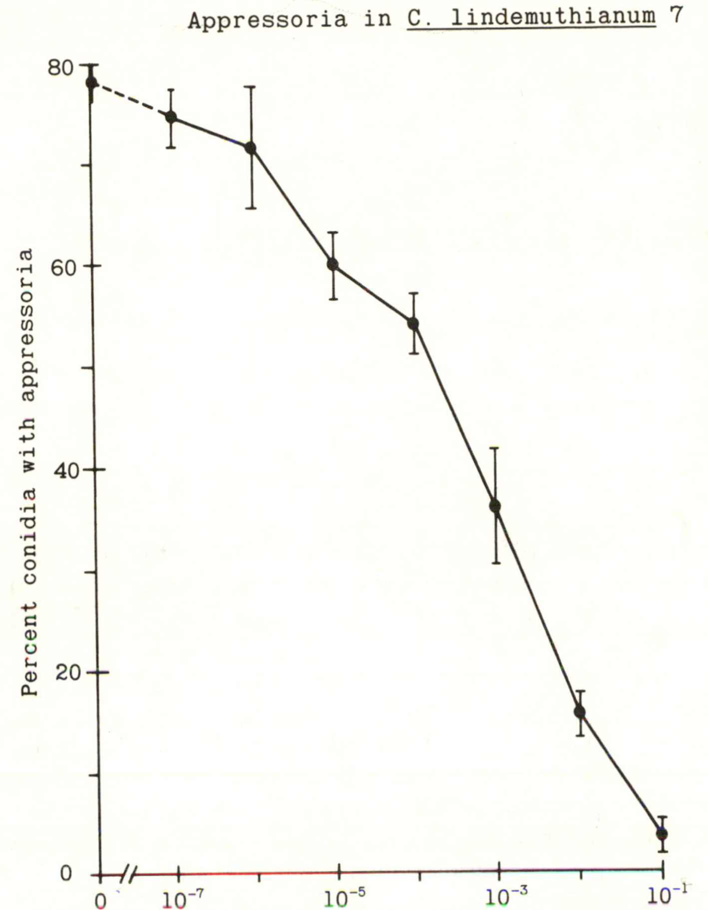

FIGURE 1. Germination morphology of spores of Colletotrichum lindemuthianum incubated in solutions of varying sucrose concentration. The number of spores with appressoria is expressed as a percentage of the total number of germinated spores (with appressoria and with only germ tubes). Standard error bars are shown for the four replicate experiments.

Include enough information to make graph understandable on its own. Plot independent variable on *X*-axis, dependent on *Y*-axis, clearly labeled. Break *X*-axis to save space. Standard error bars indicate reliability of each sample mean as indicator of population mean.

Depending on instructor's preference, figure may appear with legend close to its first mention or, as in papers for publication, legends may be typed on separate page and grouped with figures at end of paper.

CHAPTER 2

Writing a Review Paper

A scientific review paper is a critical synthesis of the research on a particular topic. Biologists read review papers to stay abreast of the current knowledge in a field and to learn about subjects that are unfamiliar to them.

When you are assigned to write a review paper for a biology course, you need to rely on many of the same strategies used by authors of published reviews. You are expected to address readers similar to yourself and your classmates — people with a background in the broad subject area but without specialized knowledge of your particular topic. Your aim is not only to inform, but also to evaluate and interpret. A good review still bears the stamp of the writer's own thought processes.

CHOOSING A TOPIC

Do not underestimate the importance of choosing a suitable topic. Ideally, it should be (1) interesting to you; (2) not so broad that it is unmanageable; (3) not so narrow that you can't find enough information on it; (4) not so difficult that you can't fully understand it.

A common approach is to start with a broad topic and do some general reading about it, gradually narrowing it down to a workable size. Suppose you decide to write a paper about orchids. Because orchids are a large, diverse group and much has been written about them, you will need to restrict yourself to some specific aspect of these plants. For example, you might decide to write about the unusual interactions between orchids and their insect pollinators. Eventually you might narrow your topic even further, perhaps confining yourself to only a few species or to several, similar kinds of pollination interactions.

As you narrow your topic and become familiar with the literature, you need to develop a sense of your main objectives. What question will your paper address and from what perspective? Are you shifting toward a particular viewpoint or conclusion that can serve as a main point, or thesis, for the paper? If you limit your scope and define your goals early in the project, your reading and note-taking will be more directed and your time will be more productive.

Occasionally the first topic you choose may be *too* narrow and you will have to enlarge it, or shift to a different subject altogether. Allow time for false starts, delays, and topic changes. Recognizing a good thing to write about requires thought and effort.

SEARCHING THE LITERATURE

■ Start with general references.

Before plunging into the technical literature, make sure you have a solid background. Use the library *card catalog* to locate textbooks or other references likely to include a general discussion of your topic. These will summarize the broad subject area, clarify key terms and concepts, and give you a feeling for the kinds of questions scientists have been investigating. Most texts also list additional readings on more specialized topics. Finding books through the card catalog will be frustrating if your topic does not seem to be listed. Be persistent; you may not be using the correct headings. A useful reference here is *The Library of Congress Subject Headings* (ninth edition), published by the U.S. Library of Congress, Washington, D.C. (1980). This work contains standardized Library of Congress headings for a large number of topics. Looking up your subject here will guide you to terms under which you will be more likely to find information.

Many libraries have adopted a computerized "card" catalog system, which has greatly facilitated the search process.

■ Use scientific abstracts and indexes.

These are invaluable aids in locating technical information on your topic. Both indexes and abstracts list papers published in a wide variety of journals; abstracts also provide short summaries of the papers. Each reference work has its own method of organization, which is easily understood if you spend a little time reading the instructions section for users. The effort is well worth it. The *General Science Index* and the *Biological and Agricultural Index* are good places to start if you are working with a broad topic you need to narrow down, and if you have not had much experience with scientific literature.

General Science Index. This source indexes subjects in diverse scientific fields, including botany, zoology, physiology, microbiology, conservation, oceanography, nutrition, astronomy, and earth sciences. It lists papers in selected technical journals (for example, *Ecology*, *Evolution*, *Nature*, *Science*) as well as many nontechnical magazines (*Natural History*, *American Scientist*, *Scientific American*, *Audubon*, *Science Digest*, and so on). Nontechnical articles are good places to start if you have little background knowledge of your subject. They give an overview of the topic, list other useful references, and prepare you for the more specialized language of technical journals — which ultimately should be your major reading material. Many of these nontechnical articles are written by scientists who have also published technical papers on the same topic; later in your search, you may wish to locate these papers, too, by looking up the authors' names in another index such as *Biological Abstracts* and *Science Citation Index*.

Biological and Agricultural Index. This source is particularly useful for topics in environmental and conservation sciences, agriculture, veterinary medicine, and related areas of applied biology. It lists articles in selected biological journals (for example, *Heredity*, *American Naturalist*, and *Ecological Monographs*) as well as in more specialized periodicals such as *Crop Science*, *Tropical Agriculture*, *Water Management*, *Animal Production*, and *American Veterinary Medical Association Journal*. Beginners find it especially useful because it includes many periodicals devoted entirely to reviews. If you can locate a review paper dealing with your general subject area, you will have a better grasp of current research in the field and may pick up some clues about narrowing your topic productively. The Literature Cited section of a review paper is an excellent source of further information (see p. 42). Moreover, reading a professional review paper helps prepare you to write a review of your own.

Biological Abstracts. This reference is widely used by professional biologists. It includes five different kinds of indexes: an author index; a generic index (using the scientific names of organisms); a biosystematic index (using taxonomic groupings from the phylum level through the family); a concept index (using broad subject headings); and a subject index (based on more specific topics — key words in titles — and including the common names of organisms). These indexes list relevant papers by reference numbers. Looking up each number in an accompanying volume gives you full bibliographic information on the paper plus a short summary of its contents. If the paper sounds relevant to your topic, you can track it down and read the whole work.

Suppose, for example, you are researching a paper on cannibalism in fish. Looking up the key word, *cannibalism,* in the most recent subject index, you find eighteen different reference numbers. Each number is preceded by the title (or part of the title) of the paper to which it refers (Fig. 1). Of these, the paper numbered 22252 ("Cannibalism in fresh water fish . . .") sounds most promising. The next step is to look up number 22252, read the abstract of the paper (Fig. 2), and decide if it is relevant to your topic. In this case the abstract suggests the paper contains useful information, and you would be wise to look at the paper. In practice, you may need to skim the abstracts of many papers listed under a particular key word, because it is not always possible to gauge a paper's contents from the portions of titles appearing in the subject index. Remember, too, that you will need to define the scope of your literature search by deciding how far back in time to search for papers on your topic.

Science Citation Index. If you already know of one paper relevant to your topic, you can use this work to locate other, more recent papers by authors who have mentioned (or *cited*) it. Chances are that those authors have written about the same subject. First, look up the known paper (say it's by J. L. Jones in 1970) under the author's name in the *Citation Index*, starting with the most recent volume and working backwards. You may find one or more references to other authors who have cited Jones (1970) in their own works. Full bibliographic information on these papers will be given in an accompanying volume, the *Source Index*. You can then go to the appropriate journals for the papers themselves.

There are several other ways to use the *Science Citation Index*. For example, if you suspect that J. L. Jones has recently published other papers on the same topic, you can use the Source Index on its own to find these, listed under Jones's name. If you know that scientists at Northwestern Medical School in Chicago are particularly interested in the contraceptive effects of breastfeeding, you can use the *Corporate Index*, which lists articles all originating at the same institution. Finally, if you have thought of one or more key words pertaining to your topic, you can look each one up in the *Permuterm Subject Index*, which lists *pairs* of key terms (some of which may be relevant to your research) along with recent articles whose titles include both words in the pair.

JUVENILE ADULT INJURY	**CANNIBALISM**	OBSERVATIONS OF INTRA	20327
ON DENSITY CLUTCH SIZE		THE BIOLOGY OF THE KES	95322
N CETACEAN PREDATION		THE STATUS OF THE CUT	2203
ATION INHERITANCE OF		BEHAVIOR IN NOCTUID LA	75106
ANORIDAE GROWTH RATE		COMPETITION DENSITY DE	84359
CALLING BEHAVIOR EGG		EMBRYONIC DEVELOPMEN	116962
ZOAN FUNGI PREDATION		FEEDING WHITE-TAIL DISEA	22507
PERATURE PHYSIOLOGY		IN FRESH WATER FISH HET	22252
E DYE INDUSTRY ISRAEL		IN MURICID SNAILS AS A P	27511
APTIVE SIGNIFICANCE OF		IN STICKLEBACKS GASTER	95414
CTORS INFLUENCING EGG		IN THE LAND SNAIL ARIANT	22282
RODUCTIVE BENEFITS OF		IN THE MOSQUITOFISH GA	105988
DESTRUCTIVE PARENTAL		IN THE PRIMITIVE ANT AM	22246
NOSTOMATIDAE CHICKEN		LIFE CYCLE SCANNING EL	118962
AEDES-AEGYPTI LARVAE		MORTALITY IMPORTANCE	22582
ECAPOD DIEL MIGRATION		SEASON JUVENILE ADULT	32447
ENILE SEASON BEHAVIOR		SPAWNING FOOD AND FE	32275
PULATION COMPETITION		TEMPERATURE DEVELOPM	52677

FIGURE 1. Entry from subject index of *Biological Abstracts*.

22252. MEHROTRA, B. K. and KUM KUM HATHI. (Zool. Dep., Jodhpur Univ., Jodhpur-342001, India.) ZOOL ORIENT 2(1/2): 69–70. 1985[recd. 1986]. **Cannibalism in fresh water fish *Heteropneustes fossilis*.**—Cannibalism in *H. Fossilis* was reported during the Laboratory studies of this fish. At a temperature of 18–25° C cannibalism occurred during fasting when the fishes were being prepared for physiological experiments.

FIGURE 2. Abstract from *Biological Abstracts*.

Current Contents. This source indexes recent articles in a variety of life sciences, including biology, agriculture, and medicine, by reproducing the tables of contents of numerous journals. If you have found one or more journals that tend to publish papers on your topic, then *Current Contents* gives you a convenient way to keep up to date on the latest literature.

This reference also includes authors' addresses, enabling you to contact an author directly to request a copy of the paper if the journal is unavailable to you. Authors usually furnish reprints gladly and without charge as long as their supply lasts. You can also order a photocopy of the paper through the interlibrary loan system of your library. It may take as long as several weeks to obtain sources using either method, so allow plenty of time.

The following are other useful references for papers in biology and related fields:

Aquatic Sciences and Fisheries Abstracts
Chemical Abstracts
Environment Index
Index Medicus
International Nursing Index
International Pharmaceutical Abstracts
Oceanic Abstracts
Pollution Abstracts
Psychological Abstracts
Wildlife Review
Zoological Record

■ Use the Literature Cited section of relevant papers to find additional sources.

One of the most efficient ways to expand your list of references is to take advantage of the experience of established authors. Each time you find a useful paper on your topic, read its Literature Cited section carefully. You may discover many titles that are relevant to your search. The author's comments about these works in the Introduction or Discussion of the paper may give you further clues about their contents, perspective, or significance.

■ Consider a computer search.

Most libraries offer computerized searches of online data bases such as BIOSIS (BioScience Information Service), AGRICOLA (Agricultural Online Access), CAB (Commonwealth Agricultural Bureau), and MEDLARS (Medical Literature Analysis and Retrieval System). Computer searches are usually done by a reference librarian. You provide several key words or search terms pertinent to your topic, and within several minutes the computer provides a list of relevant references. Some search services also supply abstracts or give you the opportunity to order the articles themselves via computer terminal.

A computer search strategy must be planned carefully. If the key words are too broad or if a particular word has more than one meaning, the computer may retrieve hundreds of titles, of which only a few will be useful. If the key words are relatively specific, you may get a high proportion of useful sources but miss some that give a broader view of the topic. Also remember that no computer search can give you every possible reference. There is always the possibility that some useful information may be hidden in an obscure book or journal. Only a thorough manual search, combined with good luck, may lead you to such a reference. Moreover, existing data bases contain only relatively recent literature. For topics that require a historical approach or lean heavily on literature prior to the 1960s, a computer search alone will be inadequate.

■ Record full and accurate information about your sources.

Keep a master list of all the references you used. Some people do this on whole sheets of paper; others list each source separately on an index card, so that when it is time to assemble the Literature Cited section, the cards can be easily shuffled and arranged in the proper order.

Learn the kinds of bibliographic information you will need to report for each kind of reference you use (see Chapter 4). If you are not sure whether certain information is essential, write it down anyway. It is easier to omit unneeded material when you eventually type your references than to spend time searching for missing publication dates or page numbers.

Photocopy all articles you think will be important sources for your paper. Even if you take good notes, you probably will need to reread part or all of the key sources on your subject before you finish researching your topic. If an article seems only peripherally related, or if you are not sure about its relevance, you may wish to photocopy only the Abstract. Jotting down a few notes about the paper's contents or perspective will allow you to return to it later if necessary. Always write full information about the source directly on the photocopy so that you are never in doubt about its origin.

■ Be discriminating in your use of sources.

It is important to understand the distinction between *primary* sources (reports of original findings or ideas) and *secondary* sources (review articles or books based on primary references). As a review, your own paper should rely mostly on the primary literature, that is, on research papers in biological journals or in edited collections of articles. This means that although you should use encyclopedias, articles for the lay reader, lab manuals, and textbooks as sources of background knowledge, they should not be major sources of information for your paper. Cite such sources rarely, *if at all*. The same applies to reviews in scientific journals; these are invaluable summaries and introductions, but they still report knowledge secondhand. If they mention research findings that sound relevant, look up the original articles and read them yourself. The quality of your review paper depends, in part, on the sophistication of your literature search and on your ability to synthesize and interpret primary sources in your own way.

TAKING NOTES

■ Avoid plagiarism: take notes in your own words.

Plagiarism is the theft of someone else's words, work, or ideas. It includes such acts as (1) turning in a friend's paper and saying it is yours; (2) using another person's data or ideas without acknowledgment; (3) copying an author's exact words and putting them in your paper without quotation marks; and (4) using wording that is very similar to that of the original source, but passing it off as entirely your own.

This last example of plagiarism is probably the most common one in student writing. Here is an example.

ORIGINAL PASSAGE

A very virulent isolate of *Alternaria mali,* the incitant of apple blotch, was found to produce two major host-specific toxins (HSTs) and five minor ones in liquid culture. The minor toxins were less active than the major ones, but were still specifically toxic to the plants which are susceptible to the pathogen.
(Kohmoto, Kahn, *et al.* 1976, p. 141)

PLAGIARIZED PASSAGE

Kohmoto, Kahn, <u>et al.</u> (1976) found that a very virulent isolate of <u>Alternaria mali</u>,

```
the incitant of apple blotch disease,
produced two main host-specific toxins, as
well as five minor ones in liquid culture.
Although the minor toxins were less active
than the major ones, they were still spe-
cifically toxic to the susceptible plants.
```

Although the writer has altered a few words here and there, the second passage is strikingly similar to the original. *It is still plagiarism if you use an author's key phrases or sentence structure in a way that implies they are your own, even if you cite the source.* The only way to make this passage "legal" as it now stands is to enclose everything retained from the original wording in quotation marks. Better yet, the writer should put the whole passage in his or her own words and word order.

Plagiarism of this kind is usually unintentional, the result of poor note-taking and an incomplete understanding of the ethics of research and writing. Typically the problem arises when you lean heavily on notes that consist of undigested passages copied or half-copied from the original source. These become the source of all the information and ideas for your paper. When you sit down to write the first draft, it is all too easy for this material to end up barely changed as the backbone of your paper. Thus, your text becomes an amalgamation of other people's words disguised as your own. Even if you cite references for the facts and ideas, you are still guilty of plagiarism because the wording is not completely yours.

Another problem with this kind of note-taking is that it consists of reading without thinking. It allows you to speed through a stack of references without necessarily understanding the material. It conflicts with your major purpose in writing a review paper: to evaluate and interpret information on a subject. You need to start making judgments, comparisons, and contrasts while you are still working with the original sources; otherwise, your prose is just a mosaic of other people's material. Your paper, like good published reviews, should be more than just a sum of its parts.

Form the habit of taking notes mainly in your *own* words. If you are not used to doing this, you may be frustrated by the additional time it takes. However, once you start the first draft, these notes will save you much time and effort. You will have already worked through difficult material, weeded out many inconsistencies, responded to the conclusions of other authors, and made connections among related ideas. Much of the preliminary work will have already been done.

To take notes effectively you need to understand how to *summarize* and *paraphrase* material. A summary expresses the important facts and ideas in fewer words than the original; for example, the abstract of a research paper is a summary. A paraphrase expresses certain facts or ideas in different wording — your own — but usually in about the same number of words as the original. Both require that you understand the material fully before you write about it.

■ Use an orderly system.

A common method is to use index cards, putting one idea or group of related ideas from a single source on each card. The cards thus contain manageable units of information and can be shuffled around at will as you organize your paper. However, such a method can be bulky and cumbersome, and many people feel constrained by the small size of the cards. Scientific topics often require longer, more detailed notes that cannot fit on index cards.

An alternative method is to take notes on whole sheets of paper, writing on just one side so that you can cut, paste, and arrange notes later as you prepare the first draft. The backs of computer printouts are excellent for note-taking; their large size allows you to add comments or other additional notes in the margins as you go along.

Obviously, you need not take notes in complete sentences. In fact, if you try to restrict yourself to succinct phrases, you'll be even less likely to reproduce the exact wording of the original. If the author's own words *are* indispensable, enclose them within quotation marks along with the page number of the source. Do this for entire passages you wish to preserve, as well as for key words or phrases mixed in with your own notes:

> J. concludes that "despite the predictive power and elegance" of the scientific method, it can give us only a "rough approximation" of what the natural world is like (Johnson 1933, p.4).

You also need a foolproof method to distinguish between an author's ideas and your own. For example, you might use a yellow marker to highlight your ideas, or put your initials, the word *me*, or some distinctive symbol in front of any speculations and conclusions that are strictly your own.

> B. suggests that light availability is the most important factor here. (me) What about moisture requirements? Not discussed.

■ Be selective.

You will waste time and effort if you take copious notes on every source you encounter. Read first; take notes only when you have decided that the reference may be useful. Start with a short summary of the author's most important findings. Once you have narrowed your topic, you can return to the most pertinent material and paraphrase where necessary. Resist the temptation to copy out the Abstract of a paper. You will accomplish more with your own words, and you can always photocopy the Abstract (or the whole paper) to consult later, if necessary.

If you are not used to looking at biological journals, you may find research papers hard to read and understand. You may vacillate between taking notes on everything and taking none at all. Here are some guidelines to help you get the most out of a scientific paper.

1. Read the Abstract first. This will give you an overview of the study and help you decide whether to read the rest of the paper. Don't feel intimidated by abstracts containing unfamiliar terms or ideas. Often the Introduction and Discussion sections of the same papers are easier to understand.
2. If the paper seems relevant, read the Introduction carefully. Be sure you understand *why* the author conducted the research. What were the major hypotheses or predictions? Authors generally end the Introduction with a brief statement of their objectives. How are these relevant to your objectives? Once you start reading a scientific paper carefully, you need to be sure of your rationale for including this particular reference in your paper. Do not take notes on everything; instead, build on the notes you already have from other sources. Organize your reading and your thinking around the specific aspects of this paper that relate to the direction your research is taking.
3. Skim the Materials and Methods. Unless your paper involves a close examination of the methodology in a particular field, you need not understand an author's procedures in detail. Instead, try to summarize the methods in a few sentences.
4. Read the Results carefully. Pay particular attention to the author's accompanying remarks about figures or tables; these will help you understand his or her reason for including them. Do not panic if you don't uderstand the quantitative details. Focus first on the major qualitative findings. Authors often summarize these in topic sentences (see p. 97). Remember to be selective. Do not paraphrase the whole Results section; instead, summarize the findings that are most relevant to your paper.
5. Pay particular attention to the Discussion. Understand the author's argument. How do the data support his or her conclusions? Are these conclusions in accord with other work in the field? What does the author imply is the major contribution of this study? How can you use this paper as support for the point *you* intend to convey?

In summary, when you work with biological literature do not get sidetracked. It is easy to feel overwhelmed by the specifics of each author's study and lose sight of the broad picture. Keep your own paper in mind. Get a general grasp of each author's research; then take more detailed notes on whatever aspects are relevant to your objectives. Finally, remember that the importance of a particular article may not be immediately apparent. You may need to skim through many papers at first to get your bearings and return later to those that are most central to your topic.

PRESENTING YOUR MATERIAL

■ Sketch out a rough plan or make an outline.

Biological review papers are not as standardized in their format as research papers; their organization depends largely on the subject and the writer's objectives. However, most reviews have an Introduction, a body (not labeled as such, but often with headings and subheadings), a Conclusions section, and a Literature Cited section. Some have a Summary at the end. Many start with a short Table of Contents that functions as an outline to the paper. Some have an Abstract, which summarizes the major points covered and states the scope and purpose of the review.

The best way to familiarize yourself with the structure of a review paper is to look through several on topics in your general field. Many journals, such as *The Biological Bulletin* and *American Midland Naturalist*, publish review articles in addition to research papers. Other journals specialize entirely in reviews; these include *The Botanical Review*, *Physiological Reviews*, *Psychological Review*, *The Quarterly Review of Biology*, and various annually published volumes of reviews, such as *Annual Review of Ecology and Systematics*, *Annual Review of Microbiology*, and *Annual Review of Genetics*.

Before you plunge into the first draft, you will need a tentative plan. This can range all the way from a rough sketch of the order of topics to a formal, detailed outline. Some people feel hemmed in by outlines and prefer to do much of their organizing as they work out the first draft. (These are the writers who, if ever asked to produce an outline, do so *after* they have written at least one draft of the paper.) Other people depend on a fixed plan right from the beginning to organize their thoughts. There is no one correct way to plan a review paper. You must decide what kind of organizational method suits your own writing style.

At some point in the writing, you can use your outline or plan to sketch out the Table of Contents, which in long review papers is an enormous aid to the reader. (Short reviews can do without a Table of Contents as long as the material is still well organized.) Here is the Table of Contents for a student review paper in animal behavior:

THE ADAPTIVE SIGNIFICANCE OF ALARM CALLS IN MAMMALS

Note the use of headings and subheadings in this paper to give order to an otherwise unwieldy amount of material. Subdividing the text may also make the paper easier to write because you can tackle one chunk of material at a time. Be sure to use a consistent method in the text to designate the various sections — for example, major headings can be typed in capitals and centered on the page; subheadings can be put in boldface and typed flush with the left margin.

■ Introduce the subject, explain your rationale, and state your central question, objectives, or thesis.

These three tasks need to be accomplished in the Introduction. As in a research paper, an effective strategy is to start with broad statements, explanations, and definitions that orient and educate the reader. Then work down to more specific issues. Why is this subject important? What approach have you taken? Will you be giving a comprehensive summary or one that is more limited? End the introduction with a clear statement of the question you will address or the main point you wish to convey to the reader. Your objectives may be fairly specific — for example, to show that certain kinds of early childhood experiences predispose adolescent girls toward anorexia. Or you may wish to assess the current state of research on a particular problem, for instance, current treatment methods for AIDS, with the aim of making predictions about the next decade.

The length of the Introduction depends on your subject and the kind of coverage you plan to give it. An 8-to-10-page review might need only a single succinct paragraph to introduce it. A longer paper might require a whole page. Generally, one or two paragraphs are appropriate. Readers will soon become confused if you do not tell them your objectives fairly early in the paper.

Here is the Introduction from the first draft of a term paper for a molecular biology course. The numbers in parentheses refer to sources in the Literature Cited section (see discussion in Chapter 4). How might this student paragraph be improved?

> Many different articles were read about the molecular genetics of human growth hormone. This paper will focus especially on hGH deficiencies. Human growth hormone (hGH) is a polypeptide hormone, produced from within a gene cluster on chromosome 17, that controls much of the physical growth of the infant and child (1, 2). Since time is limited, this paper cannot cover all possible aspects of hGH, so a narrower approach has been taken.

In this Introduction two sentences, the first and the last, say nothing essential. The reader assumes you have read about your topic, and you can easily show how you plan to narrow the focus by stating exactly what this focus is. Remember that the Introduction gives you your first chance (perhaps your only chance) to interest the reader. Obviously your instructor must read the paper whether he or she wants to or not, but if you get off to a forceful and interesting start, the paper will have a much better effect.

A second problem with the paragraph is that we don't get a clear sense of the writer's *specific* purpose or rationale. We know that the paper will focus mainly on hGH deficiencies, but we do not know how or why. If the writer fails to portray the subject as important or intriguing, it is difficult for the reader to feel it is.

Here is the second-draft version of the introduction. Notice how the writer has omitted the unnecessary sentences and filled in the "gaps" by expanding the more important parts of the original. The revised Introduction conveys a clearer idea of what this paper is about and why this subject is interesting.

> Human growth hormone (hGH) is a polypeptide hormone, produced from within a gene cluster on chromosome 17, that controls much of the physical growth of the infant and child (1, 2). Deficiency of hGH, a heritable disorder, can result in infantile dwarfism and retardation (3, 4, 5). New research methods, including recombinant DNA technology, have made it possible to determine the molecular basis of such

deficiencies. In this paper, I will summarize current knowledge of the molecular genetics of hGH and suggest ways in which continued research may help physicians treat infants with a deficiency of this hormone.

■ Build a focused discussion.

Many student review papers are little more than summaries, boring ones in which the writer has retreated from the reader's sight. It is not enough just to regurgitate the contents of a series of papers one by one. You need to *relate* this material to your principal objectives. Present your information selectively, and use it to support or illustrate the statements you wish to make. A good review interprets the literature from the writer's own informed perspective and gives the reader a sense of integration, development, and focus.

The following paragraph is from a paper on the adaptive value of cannibalism in animals. Notice how the author uses examples from the literature to illustrate and develop the generalization in the first sentence.

Following the reasoning of West Eberhard (1975), we may predict that cannibalism may be more likely when the potential victims are unusually vulnerable and easily obtained as food. Such individuals are, in fact, the predominant victims of cannibalism in many species. For example, most cannibalism in Tribolium is performed by larvae and adults on the defenseless eggs and pupae (Mertz and Davies 1968). Pupae are also eaten by larvae in caddisflies (Gallepp 1974), and injured or weak immature stages are devoured by older nestmates in many species of social ants (Wilson 1971). In crows (Corvus corone), cannibalism of eggs and nestlings by intruding adults is more frequent when the parents are absent from the nest, leaving their young more vulnerable to attack (Yom-Tov 1974).

The next passage is from a paper for a plant pathology course. It discusses host-specific toxins, substances produced by pathogenic (disease-producing) fungi that attack certain plants. Notice how the author presents selected information from the literature to critically examine a particular hypothesis. The writing conveys authority and a thorough familiarity with the material.

> Changes in the permeability of host cell membranes after being exposed to toxins suggest that these substances may bind to a receptor site in the cell membranes of susceptible hosts. Strobel (1974) claims to have found such a receptor site in the membranes of sugarcane cells treated with helminthosporoside, a toxin isolated from the pathogenic fungus <u>Helminthosporium sacchari</u>. The site contains a protein of molecular weight 48,000 daltons that specifically binds the toxin. Strobel has proposed that the binding of the toxin stimulates the activity of an adjacent membrane-bound enzyme, potassium-magnesium ATPase, which maintains ion balance across cell membranes. Such stimulation could disrupt the membrane's selectivity to ions, resulting in the characteristic symptom of electrolyte leakage.
>
> Several authors have criticized Strobel's methodology and interpretations. For example, Wheeler (1976) doubts that the preparation used for helminthosporoside assay and structure determination was sufficiently free from impurities, and he argues that some experiments lacked sufficient replication. Others have said that the toxin-binding data were not graphed correctly and that regraphing them suggests the binding activity is, at best, weak (see, for example, Hanchey and Wheeler 1979). These and other criticisms have been reviewed by Yoder (1980).

■ Document your paper thoroughly.

Whenever you refer to another author's work or ideas, cite your sources using conventional methods of literature citation (see Chapter 4). In a review paper (as in the Introduction and Discussion of a research paper), you need to cite references repeatedly. Look at the following sentence from a review by Hepler and Wayne (1985, p. 412).

> Red light triggers a large array of physiological and developmental events that require Ca^{2+}, including chloroplast rotation in *Mougeotia* (55, 78, 247–249), spore germination (254–256) and cell expansion (37) in *Onoclea*, leaflet closure in *Mimosa* (22, 23, 237), root tip adhesion in *Phaseolus* (229, 279), peroxidase secretion in *Spinacia* (113, 164, 165), membrane depolarization in *Nitella* (261), as well as activation of NAD kinase (1, 218, 232) and inhibition of mitochondrial ATPase (212).

Using many citations may seem strange at first. You may feel they impede the flow of your writing. However, readers of scientific papers are accustomed to such interruptions, and you will get used to them, too. Remember that literature citations serve an important function: they tell readers where to find additional information. Careful documentation also reflects the thoroughness of your literature search as well as your honesty in acknowledging the sources of your material.

■ Use quoted material sparingly.

Many beginning writers, unsure of their own voices or uncomfortable with the material, tend to fall back on quotations to get them through "rough spots" in the paper. Sometimes they construct whole paragraphs around a series of quotations from different authors, stringing these together with a few scattered phrases or sentences of their own. The text thus becomes a collection of other people's words:

> Studies of the Baltimore butterfly (*Euphydryas phaeton*) showed that "larvae occupied communal nests of various sizes" and "commonly cannibalized unhatched eggs in the same colony" (Monti 1950). "In both field and laboratory tests, there was a higher incidence of cannibalism by larvae occupying large colonies" (Monti 1950). Also, "cannibalistic acts occurred at a higher frequency under conditions of food shortage," when the larval foodplant, turtlehead, "was in short supply or extensively defoliated" (Mulry 1960).

In such a passage the reader loses track of the writer and the writer loses authority. The quotations do not enhance the text — they detract from it, suggesting that the writer hasn't come to terms with the material and is either too inexperienced or too lazy to use his or her own words. Biological authors rarely use quoted material, relying instead on careful, concise paraphrases or summaries; you should do the same. For example, the quotation-ridden passage above can be rephrased and condensed in the author's own words:

> Studies of the Baltimore butterfly (*Euphydryas phaeton*) showed that cannibalism of eggs by the communal larvae was more frequent in large colonies than in small ones (Monti 1950). Cannibalism also increased when the larval foodplant, turtlehead, was scarce (Mulry 1960).

When *is* it appropriate to use the exact wording of an author? Occasionally you may wish to include a quotation to establish or emphasize an important point or state a precise definition. Sometimes you may feel that an author's exact words are indispensable in conveying a particular viewpoint or idea:

> Bem (1981, p. 255) defines a *schema* as a "cognitive structure, a network of associations that organizes and guides an individual's perception."
>
> To Eiseley (1961), Darwin was "the man who saw the wrinkled hide of a disintegrating planet, glyptodonts and men, all equally flowing down the direction of time's arrow; he was a master artist and he entered sympathetically into life" (p. 351).
>
> In his controversial book, *The Naked Ape*, Morris (1967, p. 211) concludes that the survival of the human species depends on an increasing awareness of our biological heritage:
>
>> We must somehow improve in quality rather than in sheer quantity. If we do this, we can continue to progress technologically in a dramatic and exciting way without denying our evolutionary inheritance. If we do not, then our suppressed biological urges will build up and up until the dam bursts and the whole of our elaborate existence is swept away in the flood.

When you do use quoted material, cite the source in the text using either the name-and-year or number method (see Chapter 4). Some biological authors include the page number of the book or article in the citation; others do not. Be sure to reproduce the quoted material *exactly*. Introduce and punctuate quotations properly, using the following rules.

1. Do not substitute single quotation marks (' ') for double ones (" "). The former are restricted to quotations within other quotations. (In British usage, single quotation marks are used first; double quotation marks are used to set off quotations within quotations.)
2. A short quotation can be integrated into the text. Make sure it fits in grammatically with the rest of the sentence.

> Dawkins (1976, p. 206) suggests that memes, as units of cultural transmission, can replicate themselves by "leaping from brain to brain via a process which, in the broad sense, can be called imitation."

For a quotation longer than four typed lines in your paper, omit the quotation marks and use a colon to introduce it. Indent each line of the passage five spaces from the left margin, but keep it double-spaced.

3. Place periods and commas *inside* the quotation marks; semicolons and colons go *outside.*

> Jones (1987, p. 2) calls Davidson's explanation "the most exciting model of this century."
>
> Davidson's explanation is "the most exciting model of this century," according to Jones (1987, p. 2).
>
> Jones (1987, p. 2) considers Davidson's explanation "the most exciting model of this century"; unlike previous models, it gives rise to far-ranging predictions about how major evolutionary changes occur.

If exclamation marks, question marks, and dashes are part of the quoted material, put them inside the quotation marks. If they are part of your own sentence, put them outside.

4. If you need to interrupt a quotation to omit one or more words, indicate the omission by an ellipsis (three periods, separated by spaces). If the ellipsis falls at the end of a sentence, put a period after it.

> Morris (1976, p. 211) concludes that humans must "somehow improve in quality. . . . If we do this, we can continue to progress technologically in a dramatic and exciting way without denying our evolutionary inheritance."

5. If you insert clarifying or explanatory material into a quotation, put such material in brackets [ ].

> Morris (1976, p. 211) concludes: "If we do this [improve in quality], we can continue to progress technologically in a dramatic and exciting way without denying our evolutionary heritage."

■ End with general conclusions.

Despite the importance of an effective Conclusions section, this part of a review paper gets the least attention from many beginning writers. By the time they reach this point, most writers are worn out and feel they have nothing more to say. They are so anxious to be done with the paper that they make a hasty retreat, summing up the topic in a sentence or two and then coming to a weary stop.

> Thus, the role of host-specific toxins in plant disease is a complex topic. There are still many questions for scientists of the future to answer.

The following student paragraph, however, summarizes the major points in the paper and adds a final perspective:

```
In conclusion, current evidence shows that resist-
ance or susceptibility is an inborn trait of a
```

```
plant.  Some plant pathogenic fungi can produce tox-
ins that, by disturbing host physiological func-
tions, affect the onset and development of plant
disease.  However, it is not clear whether these
substances have a primary role in determining re-
sistance or susceptibility.  Perhaps some other
mechanism may determine specificity, and toxins may
exert their damaging effects secondarily.  Future
research on host-specific toxins will help to answer
this question.
```

As shown above, a Conclusions section consolidates and strengthens the relationships, patterns, and arguments you have been building in the body of the paper. However, a good Conclusion should do more than merely summarize; it must also *conclude* something. Assuming that you have adequately addressed the topic, now you must answer such questions as "So what?" or "What next?" What is the significance of everything you have just told the reader? What conflicts still need to be resolved? What research must still be done? What might we expect to happen in the future?

Do not, however, introduce *new* information that really belongs in the body of the paper. Avoid complex questions or issues that you can't fully address. You do not want to throw readers off course or leave them hanging in mid-air. On the contrary, you want to tie up loose ends and finish with a satisfying sense of closure.

There is no set length for a Conclusions section. In published reviews, it varies from a single concise paragraph to a page or more of text. The length depends on the topic, the author's aims and depth of coverage, and any length restrictions on the paper. Plan ahead as you are drafting the manuscript so that you have enough space, time, and energy to end your paper effectively.

CHAPTER 3

Using Tables and Figures

The text of scientific papers is often supplemented with tables and figures (graphs, drawings, or photographs). Such materials can convey certain types of information much more effectively than words alone. A table can help you compare the results of a variety of chemical analyses. A graph can illustrate the effect of temperature on the growth of bean seedlings. A line drawing can depict an aggressive interaction between two fish, and a photograph can record important features of your study site.

It is not true, however, that tables and figures are essential in a scientific report. Biology students sometimes think that *all* data must be tabled or graphed to produce "professional" results. Even in published papers, unnecessary tables or figures sometimes find their way into print, wasting the reader's time and raising printing costs. Your credibility as a scientist will suffer if you use tables or figures just to impress the reader, when simple text will do as well or better.

A crucial task in scientific writing is deciding on an effective way to display whatever relationships, patterns, and trends are present in the data. You need a good sense of what quantitative data you are seeking and why. What statistical analyses are most appropriate to address the question you have asked? This concern should have high priority in the early stages of your research, not just at the end when there is no time to compensate for mistakes, problems, or oversights.

A second and related concern is how to incorporate your quantitative data smoothly into the manuscript. Can your measurements and statistical analyses be summarized easily in the text, or is a table called for? Would a graph be more effective than a table? Rarely is there only one way to depict a given data set; however, one way may be superior to others. You need to review the scientific argument you are making and determine how to display the evidence most convincingly.

TABLES

■ Use a table to present many numerical values or (occasionally) to summarize or emphasize verbal material.

A large amount of quantitative information can be tedious and cumbersome to report in the text. If you put it in a table, the reader can take in everything at a glance, as well as compare one item with another (see Table 1).

Do *not* use a table when it is more important to show a pattern or trend in the data; consider a graph instead. Also, never use a table (or figure) when you could just as easily summarize the same material in the text. Look, for example, at Table 2. This table is unnecessary because it does not contain much information. Its contents can easily be put into words: "In Morrisville, Ohio, I found *Plantago major* at three out of five vacant lots sampled (Cranston Road, Poolville Road, and Hamilton Street; but not at Main Street or Maple Avenue)."

Similarly, Table 3 is unnecessary. The main message (that seeds neither imbibed water nor germinated following any of the treatments) could easily be stated in the text.

Table 4 has not progressed very far from the raw-data stage. The

TABLE 1. Soil analyses of four farm fields near Malverne, Vermont.

Site	Macroelement concentration (g per 1000 g dry soil)					
	Total N	P	K	Ca	Mg	S
Schwab farm	0.1	0.9	12.2	15.1	11.6	0.6
Toomey farm	0.8	0.7	6.6	8.3	5.4	0.5
Charles farm	2.6	1.1	17.1	15.9	13.1	1.1
Hendrick farm	2.1	0.7	18.3	26.8	13.6	0.8

TABLE 2. Occurrence of Plantago major at five vacant lots in Morrisville, Ohio.

Location of site	Present(+) or absent (-)
Cranston Road	+
Poolville Road	+
Main Street	–
Hamilton Street	+
Maple Avenue	–

results will be more meaningful after some simple calculations, allowing us to replace the whole array of numbers with two succinct sentences in the text:

> After being visually isolated from other fish for two weeks, each of 11 male *Betta splendens* was shown its reflection in a mirror and observed for 30 s. Collectively, the fish responded by approaching their images ($\bar{x}$ = 3.0 times, SD = 2.1), biting the mirror ($\bar{x}$ = 1.4 times, SD = 1.0), and erecting their gill-covers for an average of 15.5 s (SD = 4.4).

TABLE 3. Effect of germination inhibitors on Phaseolus vulgaris seeds.[a]

Inhibitor	Concn. (M)	Imbibition time (days)	Germination time (days)	% germination
Actinomycin-D	0.05	0	0	0
Coumarin	0.023	0	0	0
Thiourea	0.015	0	0	0
2,4-dinitrophenol	0.028	0	0	0

[a]80 seeds in each treatment. Seeds treated for 2 days at 40°C.

TABLE 4. Response of male fighting fish (Betta splendens) to their image in a mirror.[a]

Fish no.	Duration of gill-cover erection (s)	No. of approaches	No. of bites
1	20	6	3
2	13	5	2
3	10	2	0
4	7	1	1
5	15	1	1
6	16	5	2
7	19	3	2
8	20	1	0
9	14	0	0
10	21	5	2
11	16	4	2

[a]Prior to the experiment, fish had been visually isolated from one another for two weeks. Observation period for each fish was 30 s.

Tables need not be filled with numerical values. They may also be used (sparingly and carefully) to summarize numerous points, to summarize a review of the literature on a topic, to compare and contrast related items, or to list examples or details that would be too tedious to spell out sentence by sentence in the text (see Table 5).

■ Number tables consecutively, and make them understandable on their own.

Give each table a number in the order in which you refer to it in the text (Table 1, 2, and so on). Even if there is only one table (or figure) in the paper, assign it a number.

A table should be able to stand apart from the text and still make sense to the reader; therefore the *title* must adequately describe its contents.

TABLE 5. Selected examples of cannibalism in animals.

Cannibal	Victim	Example	Reference
Adult female	Adult male (mate)	Praying mantis	Roeder (1935) Edmunds (1975)
Adult female	Own off-spring	Guppies (*Lebistes reticulatus*)	Breder and Coates (1932)
Adult male	Offspring of other adults	Wasps (*Polistes* spp.)	Heldmann (1936)
		Lions	Schaller (1972)
Juvenile	Siblings or half-siblings	Milkweed bugs (*Oncopeltus* spp.)	Root and Chaplin (1976)
		Milkweed beetle (*Labidomera clivicollis*)	Eickwort (1973)
		Various birds of prey	Ingram (1959)
	Dead parent	Certain spiders (*Amaurobius ferox*, *Coelotes terrestris*)	Cloudsley-Thompson (1965)

Never use vague, general titles like "Field data," "Test results," or "Feeding studies." Instead, be more specific and informative:

> Average accumulations of uranium in lichens sampled at Collins Forest Preserve, Montana
>
> Indices of fear in pre-experimentally conditioned and nonconditioned mice
>
> Responses of inexperienced adult blue jays to monarch and viceroy butterflies

The title is usually written as a sentence fragment at the top of the table. (A sentence fragment is a group of words that lacks a subject, a verb, or both, or that does not express a complete thought. Table titles usually lack a verb.) By convention, only the first word of the title is capitalized. If necessary, you may add one or more full sentences after the title to give further explanatory information:

> Relative abundance of 11 cyprinid species at five lakes in Nassau County, New York, in 1985, 1986, and 1987. Fish were captured by seining on July 7–9 each year.

Footnotes also allow you to add clarifying information to a table. Unlike material appearing after the title, footnotes are less obtrusive and distracting. You can use numerals, lower-case letters, or conventional symbols (such as [*]) to mark footnotes. Put these as superscripts (such as [a]) at the appropriate places, and list the footnotes in order at the bottom of the table. Do not use more than one kind of designation for footnotes. If those in Table 1 have numerals, for example, use numerals throughout all your tables.

▪ Use a logical format.

Arrange similar elements so that they read *vertically,* not horizontally. You can see the rationale for this if you compare Table 6 with Table 7. They display the same data, but the latter is more logically set out and therefore easier to read.

Arrange the column headings in a logical order, from left to right, and list the data down each column in a logical sequence. List the data neatly under the center of each column heading, and make sure that all dashes or decimal points are aligned (see Table 7). Put a zero before each decimal with a value less than one — for example, 0.23, not .23.

▪ Make the contents concise.

A table should display selective, *important* information — not peripheral details, repetitive data, or values that are uniform and unvarying. Unnecessary data will just clutter the table and obscure more relevant data.

TABLE 6. Mean numbers and dry weights of individuals of four species of plants in 25 randomly chosen plots, each 30 x 30 cm, at a weedlot near St. James, Connecticut, August 15, 1983.

Species	Centaurea nigra	Conyza canadensis	Thlaspi arvense	Oenothera biennis
Count/plot	49.3	78.8	21.2	1.8
Dry wt/ indiv. (mg)	25.8	2.1	2.6	22.7

For example, if photoperiod, temperature, or other conditions were the same for all lettuce seedlings in your plant growth project, then these are the standard conditions, to be reported by a footnote or in the Title or the Methods section. These data do not belong in the columns of the table. Similarly, some information useful during the research may not be important enough to mention at all; for example, the reference numbers you gave to each of the mice you dissected or to a series of sampling sites. Thus, the fact that certain data take up space in your lab notebook does not automatically qualify them for a prominent position in a table.

Abbreviations are used freely in tables to save space and make the contents easier to grasp. In practice you can abbreviate many words in tables that you would not ordinarily abbreviate in the text; for example,

TABLE 7. Mean numbers and dry weights of individuals of four species of plants in 25 randomly chosen plots, each 30 x 30 cm, at a weedlot near St. James, Connecticut, August 15, 1983.

Species	Count/plot	Dry wt/indiv. (mg)
Centaurea nigra	49.3	25.8
Conyza canadensis	78.8	2.1
Thlaspi arvense	21.2	2.6
Oenothera biennis	1.8	22.7

temperature (temp), experiment (expt), and concentration (concn). Abbreviations likely to be unfamiliar to the reader (including ones you devise yourself) should be explained in footnotes. No table should contain so many abbreviations that deciphering them becomes an end in itself, demanding unlimited patience from the reader.

Do not put units of measurement after the individual values in a table. To avoid redundancy, put them in the title or the headings (see Table 7). You can also group related column headings under larger headings, so that identical details do not have to be repeated for each column. Row headings can be grouped in a similar fashion.

Make all explanatory material in tables brief and concise. Do not excessively repeat what you have already written in Materials and Methods; instead, give just enough information to refresh the reader's memory and make the table understandable in its context. If the same abbreviations or procedures apply to more than one table, give them only once. In later tables, refer the reader to your earlier footnotes: "Abbreviations as in Table 1" or "Test conditions as described in Table 4."

■ Check tables for internal consistency and agreement with the rest of the paper.

A common fault of student papers is that some tables (or figures) do not make sense in the context of the whole paper. For example, something in one table may conflict with a statement in the text; or the data in a table may seem at odds with the trend shown in a graph; or the values in a table may not make numerical sense. Be sure to carefully proofread your tables. Is each value copied accurately from your data sheets? Do all the items in a group add up to the totals shown in the table? Do all percentages add up to 100%?

FIGURES

■ Use a graph to illustrate an important pattern, trend, or relationship.

A graph is generally superior to a table when you are more concerned about the "shape" of certain data; for example, how activity in a beetle species is related to ambient temperature, or how the concentration of glucose in the bloodstream changes over several hours. Table 8 displays data that become more interesting when we put them in a line graph (see Fig. 1), because we get a clearer picture of how this population of plants changes with time.

Note in Figure 1 that the *independent* variable (time) is plotted on the *X*-axis, and the *dependent* variable (percent of plants surviving at each

TABLE 8. Survivorship of Chaenorrhinum minus from seedling emergence to production of mature seeds on railroad cinder ballast in Roslyn, New York, 1980.

Time (days)	Percent surviving	Time (days)	Percent surviving
0	100.0	56	9.9
7	72.6	63	8.3
14	55.9	70	7.5
21	38.8	77	7.1
28	25.1	84	7.0
35	18.6	91	6.8
42	14.3	98	6.7
49	11.2		

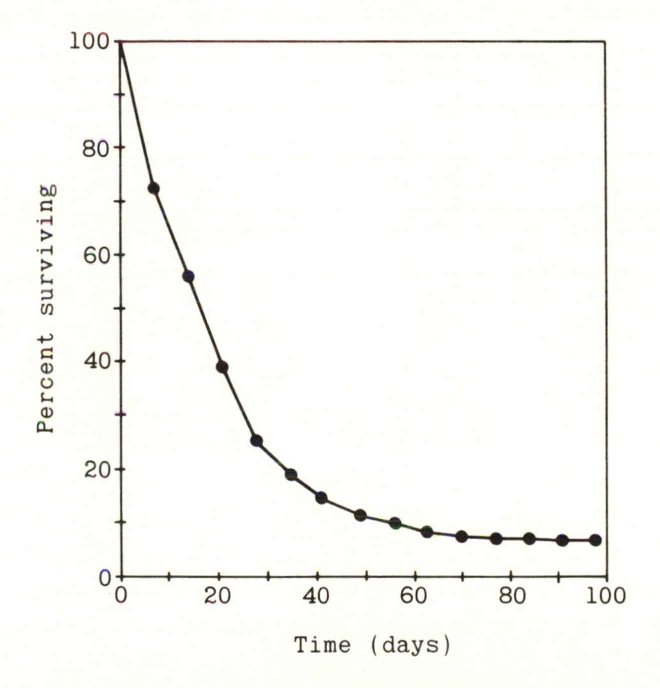

FIGURE 1. Survivorship of Chaenorrhinum minus from seedling emergence to production of mature seeds on railroad cinder ballast in Roslyn, New York, 1980.

sampling time) is plotted on the *Y*-axis. The dependent variable is so named because its value depends on (is affected by) the value of the independent variable. Another way to say this is that the dependent variable is a *function* of the independent one.

Figures 2, 3, and 4 illustrate other types of graphs commonly used by biologists. Figure 2 is a *bar graph*. The *Y*-axis depicts a numerical variable (number of flowers per plant), whereas the *X*-axis depicts qualitative (non-numerical) categories (plant species). Figure 3 is a *histogram,* or frequency distribution, showing the frequency of observations falling into a series of numerical categories plotted on the *X*-axis. Figure 4, a *scatter plot,* illustrates the correlation or strength of association between two variables. In this case, the association is between leaf length and width in a particular plant.

Generally, a graph is *not* necessary when trends or relationships are

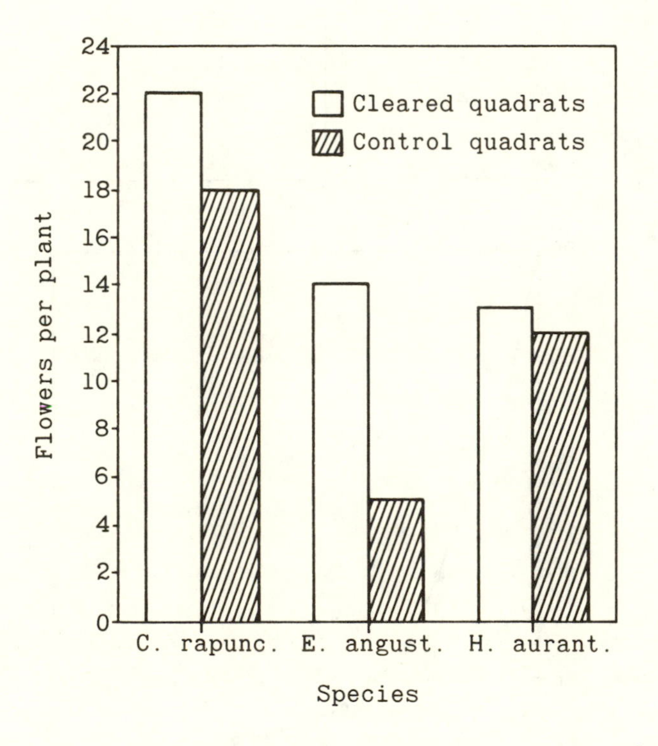

FIGURE 2. Production of flowers by three species of plants in the absence of interspecific competition (cleared quadrats) and under natural conditions (control quadrats). The plants were Campanula rapunculoides, Epilobium angustifolium, and Hieracium aurantiacum. Plotted are means for eight randomly chosen quadrats, each 1 m × 1 m.

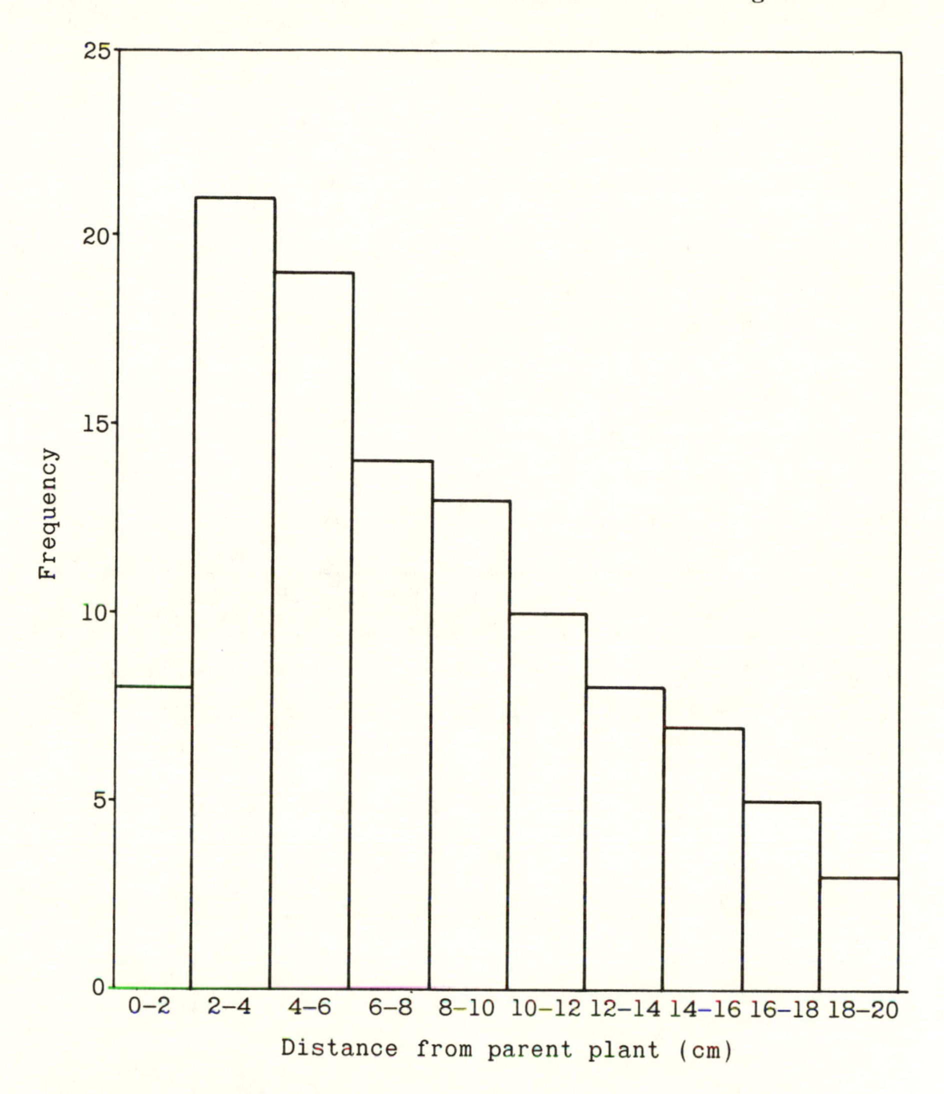

FIGURE 3. Wind dispersal of seeds from *Chaenorrhinum minus*. Data are from a 2-wk study of a single plant 15 cm tall with 35 mature capsules.

statistically insignificant (see pp. 23–24) or when data are so sparse or repetitive that they can be easily incorporated into the text. Figure 5 is an example of a useless graph because the results it portrays can be summarized in a single sentence: "Growth rates of hyphae of *Saprolegnia* and *Achlya* remained constant at 2.38 and 0.92 μm per minute, respectively, before and after treatment with penicillin."

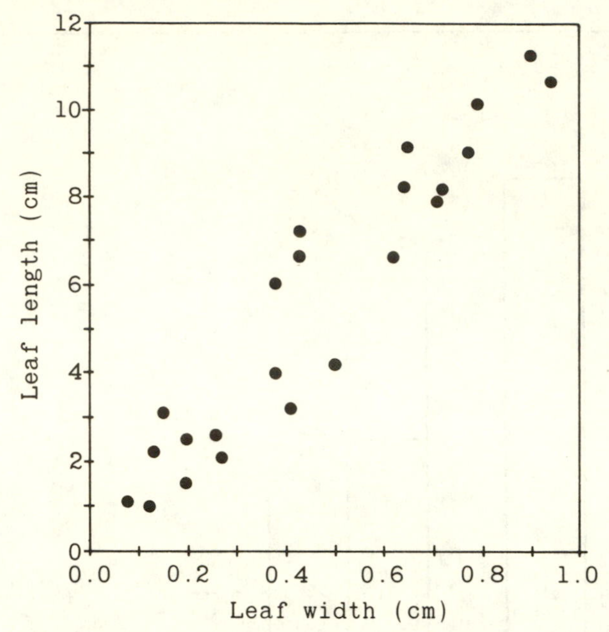

FIGURE 4. The relationship between leaf length and maximum leaf width in *Digitaria sanguinalis*. Data are from 23 randomly selected leaves from five plants. $r = 0.95$.

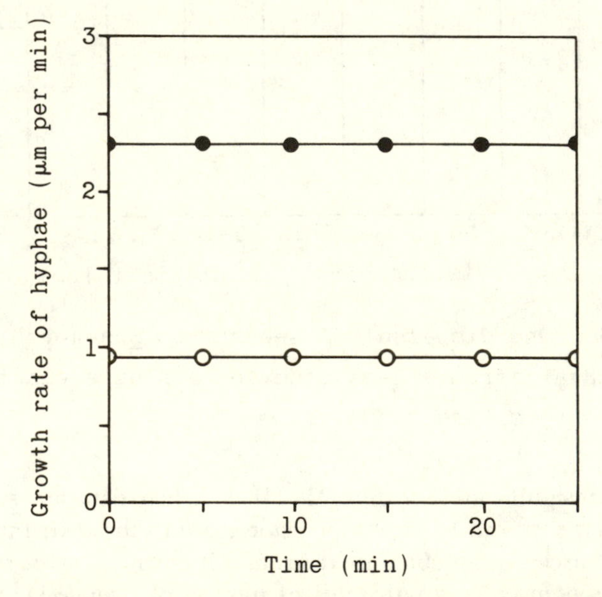

FIGURE 5. Growth rate of hyphae of the fungi Saprolegnia (filled circles) and Achlya (open circles) at 20°C. Penicillin was added to the cultures at 10 min.

■ Number graphs consecutively, separately from tables, and make them understandable on their own.

Each graph needs to be numbered in the order in which you discuss it. Use a separate series of numbers for graphs and tables — for example, Table 1, Table 2, Table 3; Figure 1, Figure 2, and so on. Use arabic numbers only, not roman numerals. Write out "Figure" in full in the *legend* (caption or title) of each graph. In the text, write "Figure" whenever it appears outside parentheses, but abbreviate it as "Fig." when you put it in parentheses. For example: "As shown in Figure 1, . . ." but "Germination rates (Fig. 1). . . ."

All figures should make sense apart from the text. The legend should be specific and informative, not vague as in "Weather conditions" or "Rainfall." Avoid simply repeating the labels of the two axes (for example, "Amount of rainfall vs time"). Instead, be more descriptive: "Rainfall fluctuations during the 1980 breeding season."

The legend is usually written as an incomplete sentence with only the first word capitalized. If necessary, you may follow it with a sentence or two of additional information, but be brief and concise (see Fig. 6). Authors submitting manuscripts to journals usually type all figure legends together on the same page, separate from the figures themselves. For a student paper, you may wish to type or print each legend on the figure itself. (See pp. 123–124 for instructions on manuscript format.)

■ Plot data accurately, clearly, and economically.

Figures submitted to professional journals must meet high standards. A sloppily prepared graph will not show off the data to best advantage, and may even cast suspicion on their quality. In student papers, too, accuracy and neatness count. Use standard graph paper, not lined notebook paper, to plot your data accurately. For a neat, professional-looking product, you can then trace this graph onto white bond paper.

Computers have greatly simplified the statistical analysis and graphing of data. For example, there are inexpensive software packages that calculate descriptive statistics such as means and standard deviations, and conduct common analyses such as t-tests and analysis of variance. More expensive professional software does more complex operations such as time series analysis and multivariate analysis of variance and regression. Most statistical software packages, whether for beginning or more advanced scientists, also produce a variety of graphs. Many of the figures in this chapter could have been produced by such software. Check to see what is available at the computer center at your institution.

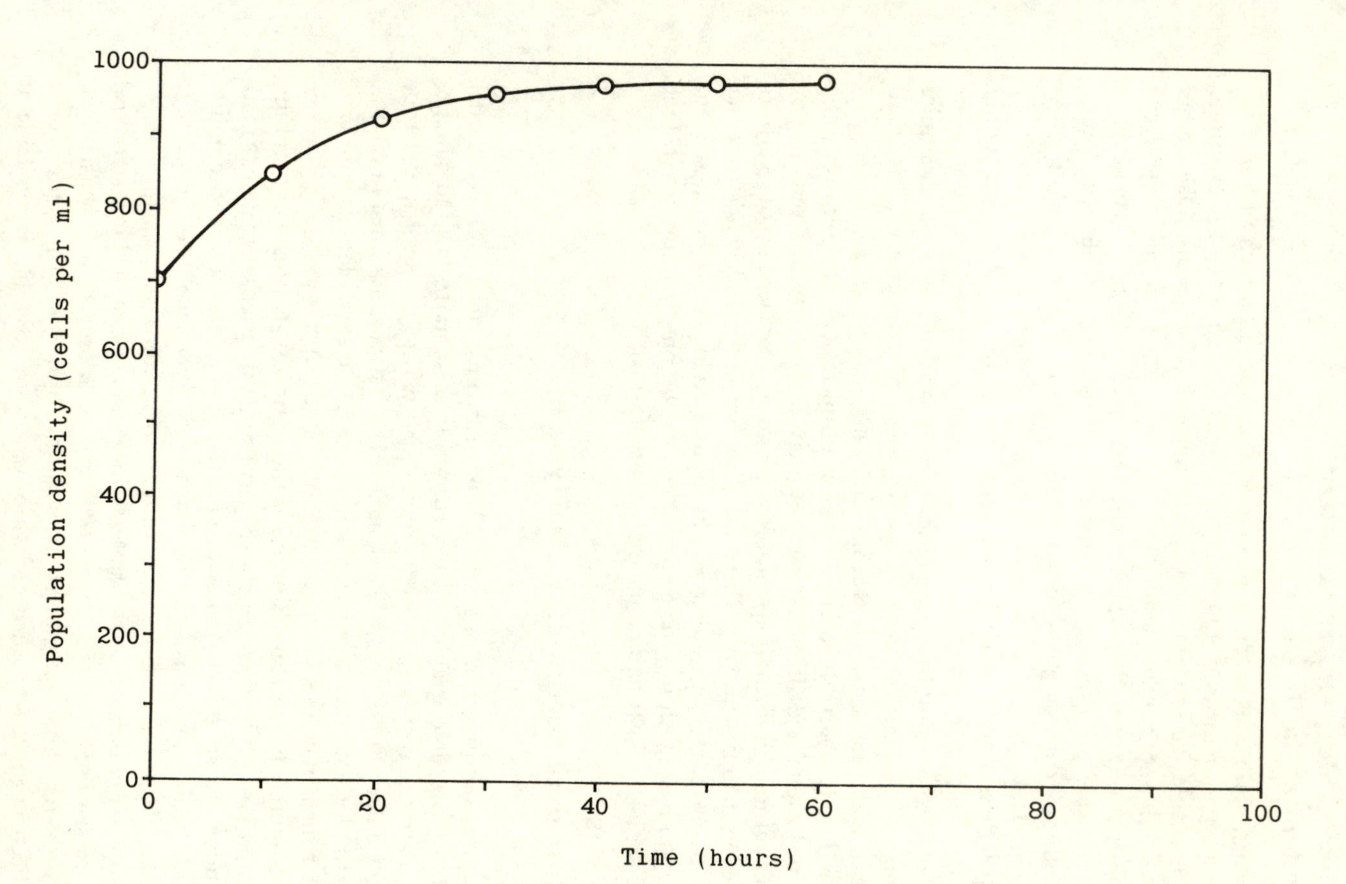

FIGURE 6. Change in population density of yeast cells growing at 27°C in malt extract broth.

Remember that the independent variable belongs by convention on the *X*-axis. Label each axis clearly using large, easy-to-read lettering. Do not forget to specify the units of measurement. Numbers should also be large and legible. You need not number every major interval along an axis; instead, use index lines to indicate the halfway points between two numbers, as in Figures 1 and 4. This kind of shorthand prevents the graph from being cluttered with unnecessary numbers and gives you more space for essential information.

Many graphs made by beginners waste large amounts of space. Look at Figure 6, for example. The *Y*-axis is shown originating at zero even though the lowest measurement is 700 cells per ml. And the *X*-axis is extended far more than is necessary because the last measurement is taken at 60 h.

Figure 7 shows the same data in a more compact format. The *Y*-axis begins with a value closest to the lowest one plotted, and the *X*-axis is only as long as needed. If you are used to seeing zero on every scale, this graph may look "wrong." However, indicating the zero mark should never be done at the expense of a graph's clarity and visual effectiveness.

An alternate way of saving space here would have been to start the *Y*-axis at zero but put a break along it, indicating that unneeded values have been omitted:

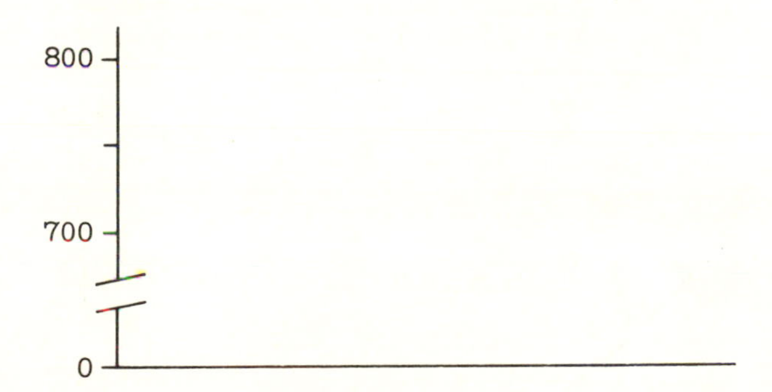

You can also save space, as well as compare two or more data sets, by putting them all in a single figure, as in Figure 8. Be sure to label each data set clearly. Explain any symbols you use in the legend or in a key placed inside the graph itself. The latter method makes efficient use of empty space. Do not pack *too* much information into a single figure, or the reader will be unable to make the necessary comparisons. A general rule of thumb is no more than four different symbols and no more than three different lines per graph.

When you plot a series of sample means, it is customary to give the reader a sense of the variability of the data or the reliability of the sample means as estimators of the population means (see p. 22). Figure 8, for

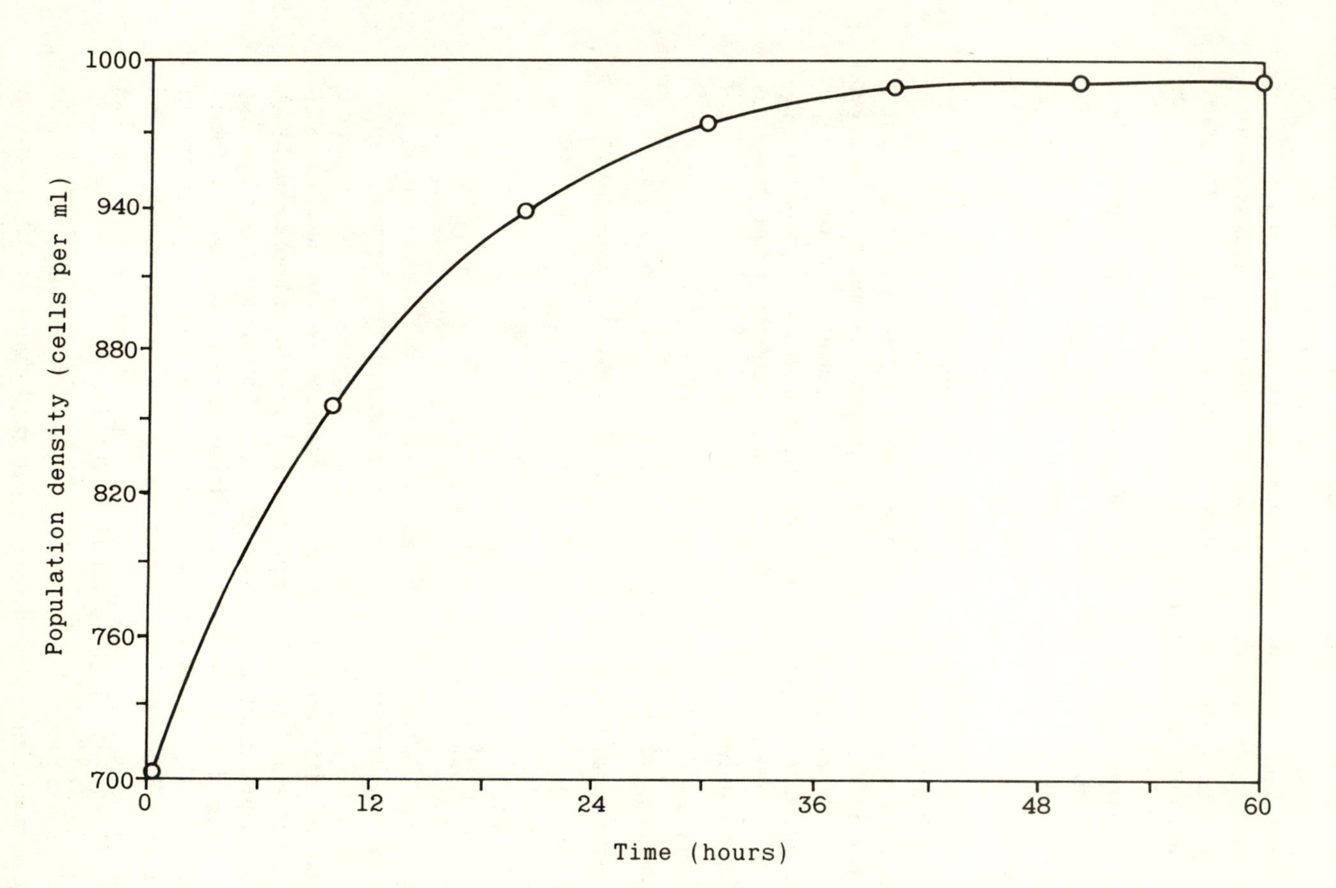

FIGURE 7. Change in population density of yeast cells growing at 27°C in malt extract broth.

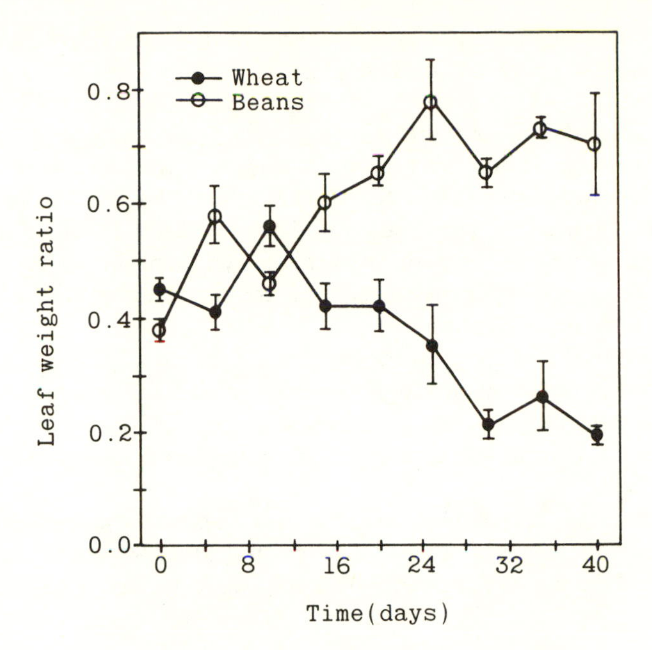

FIGURE 8. Growth rate of wheat (Triticum vulgare) and beans (Phaseolus vulgaris) as measured by the ratio of the dry weight of leaves divided by the dry weight of the whole plant. Ten plants of each species were sampled at each time. Vertical bars show the standard error of each mean.

example, shows the standard error for each mean as vertical lines extending up and down from each data point. These lines reflect the reliability of each mean as an estimator of the population mean. Using the same method to show the 95% confidence interval for each mean will convey similar information. If you wish to plot the variability of values within each sample, you could plot the standard deviation. If consideration of high and low values is important, you may prefer to show the range for each sample mean. Be sure to specify in the legend which statistics you have plotted.

■ Depict data logically, in a manner consistent with your overall hypothesis.

Remember that the data you graph may be subject to more than one interpretation. You can bring out, hide, exaggerate, or misrepresent the message carried by a given set of data by the way you choose to draw the graph. For this reason it is vital that you understand your purpose in constructing the graph in the first place.

For example, does it always make sense to connect all the dots on a graph? It depends on your data and how they were obtained and analyzed, as well as on your interpretation. In Figure 9 the data points are *not* connected; instead, the author has drawn a curve by hand to *approximate* the general downward trend suggested by the data. Most of the points are either touching or very close to this curve. Sampling error (virtually unavoidable in any experiment) accounts for the fact that all the points do not lie exactly on a smooth curve.

By contrast, in Figure 10 the author *has* joined all the points because his purpose is to show the high day-to-day variability in abundance of mayflies at the study site. We assume here that the big gaps between points are not due just to sampling variations; rather, they are meaningful fluctuations in themselves.

In Figure 11 the author has plotted a *statistically derived* (not hand-drawn) line of "best fit" to the data points. Her purpose is to depict a specific mathematical relationship between two variables. The equation for

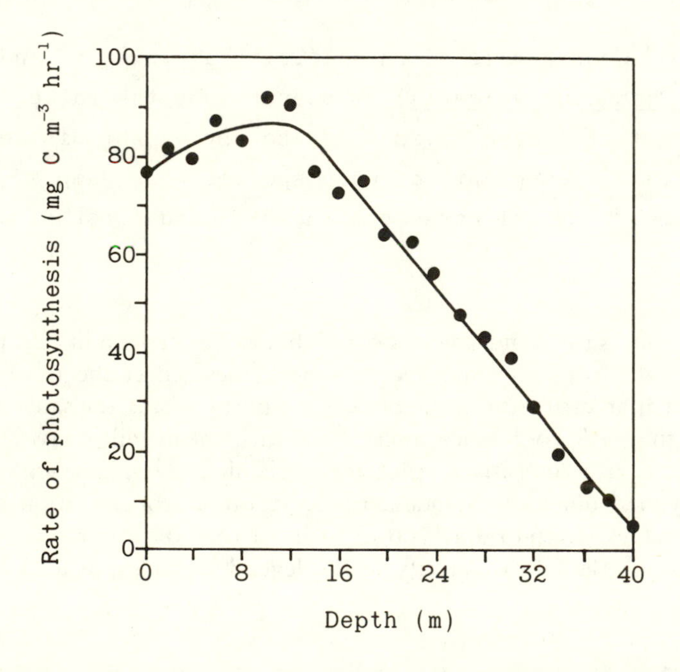

FIGURE 9. Rate of photosynthesis by marine phytoplankton as a function of depth. Data were collected at Biscayne Bay, Florida, at 1200 h, July 31, 1979.

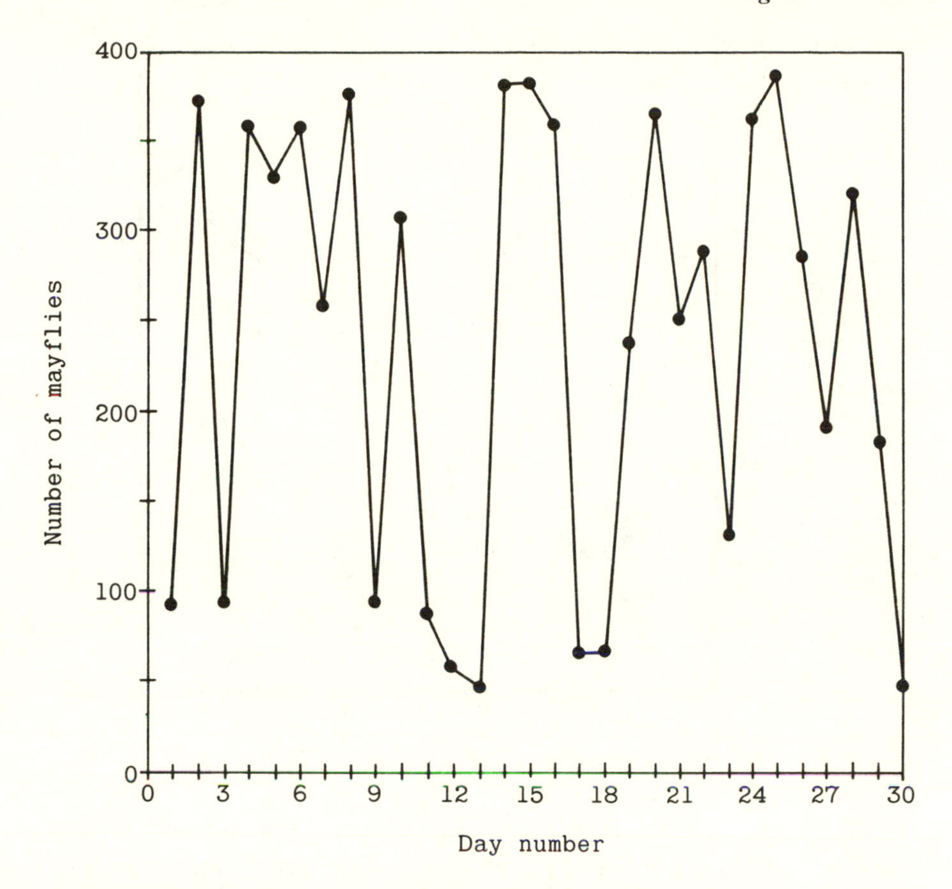

FIGURE 10. Abundance of mayflies (Ephemeroptera) at Saccho Lake, Hicksville, Minnesota, in June 1981. Data are based on daily collections at the southern shore of the pond at 1800 h.

the line, given in the legend, defines this relationship. Figure 12 illustrates a similar situation for a fitted curve. In both figures the authors have assumed that a cause-and-effect relationship exists between the two variables, that is, that values of the independent variable can be used to predict values of the dependent one. The data points give the reader a sense of the variability of the observations.

If you look back now to Figure 4 (p. 68), you will notice that the points are not accompanied by a fitted line. The author's purpose here is simply to show the interdependence of these two variables, not to suggest that the values of one are controlled or predicted by the values of the

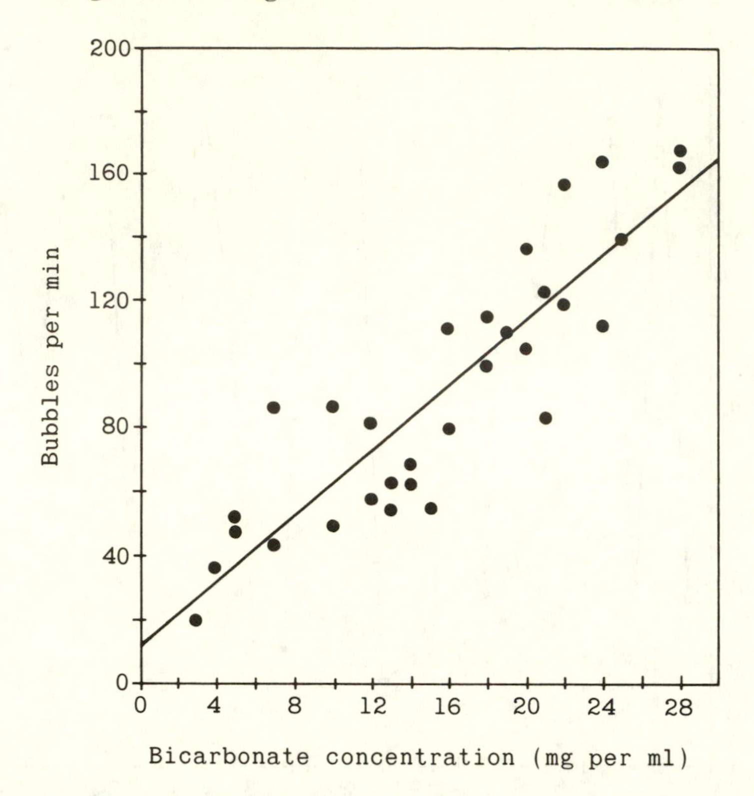

FIGURE 11. Effect of sodium bicarbonate concentration on rate of photosynthesis of Elodea canadensis, as measured by production of oxygen bubbles at 18°C. Regression equation is $y = 10.9 + 5.1x$. ($R^2 = 0.79$).

other. As in other correlation studies (see p. 24), either variable could have been plotted on either axis, since no assumptions are being made about which one is dependent and which is independent.

Thus, constructing a graph is more than just an exercise in plotting points on ruled paper. It requires that you understand the rationale behind your quantatitive methods. How is your interpretation of the data constrained by the statistical analyses you used? What assumptions are you making by describing your data in this particular way? Are these assumptions valid? Discussion of quantitative methods is beyond our scope here, but the subject is treated at length in the references listed at the end of this book.

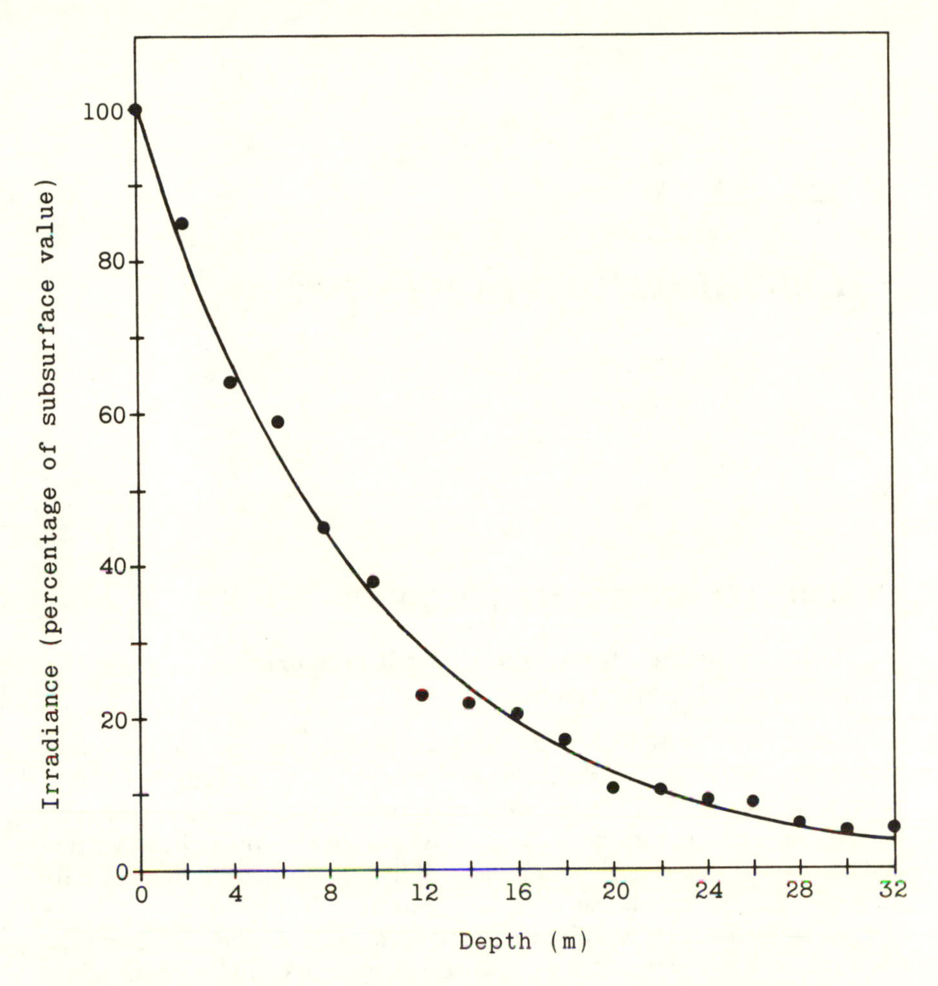

FIGURE 12. Decrease in light penetration through water. Data were collected at Glen Cove, Georgia, at 1300 h, June 22, 1983. Radiant energy (irradiance) was measured in W cm^{-2} and values at different depths were calculated as percent of the value just below the water surface. Regression equation is $\underline{y} = 101.5e^{-0.1\underline{x}}$ ($\underline{R}^2 = 0.97$).

CHAPTER 4

Documenting the Paper

CITING SOURCES IN THE TEXT

■ Acknowledge the source of all material that is not your own.

The text of a biological paper usually contains numerous *literature citations,* or references to the published studies of other authors. This is because scientists rarely work in a vacuum: hypotheses are developed, tested, and evaluated in the context of what other scientists have written and discovered. Most readers of biological literature are interested not only in the specific studies being described, but also in the bigger picture — how particular findings contribute to our current understanding of a particular scientific problem. Thus, careful *documentation,* or acknowledgment of the work of others, is essential to good scientific writing.

Biologists also need to provide literature citations because, like other writers, they have an ethical and legal obligation to give credit to others for material that is not their own. Such material includes not only direct quotations, but also findings or ideas that stem from the work of someone else.

■ Cite sources using a conventional format accepted by biologists.

Documentation formats vary from one academic field to another. Unlike writers in the humanities and social sciences, biologists rarely use footnotes or endnotes to acknowledge sources. Instead, they insert literature citations directly in the text, either by giving the last name of the author(s) and the year of publication (name-and-year method), or by referring to each source by a number (number method). Each format has its advantages and disadvantages, and certain subdisciplines of biology (and each biological journal) tend to favor one method or the other. Whichever method you choose, be consistent and meticulous in adhering to the conventions. Such rules, even if they seem arbitrary, make the reporting of references an orderly activity, minimizing confusion for writers, readers, editors, and publishers.

Name-and-Year (Harvard) Method

Cite each reference by giving the last name(s) of the author(s) followed by the year in which the material was published. A reader who is interested in, or knowledgeable about, who the particular researcher was or when the study was published will want this information.

Work by One Author

For each citation, use parentheses to enclose the name and the date.

```
The most recent study of sexual dimorphism in this
species (Jackson 1976) fails to account for. . . .
```

If the author's name appears as part of the sentence, put just the date in parentheses.

```
In Smith's (1970) model of parental investment in
fathead minnows. . . .

Black-horned locusts were first reported in Iowa by
Blum (1914).
```

Because a literature citation is considered part of the text, the period always *follows* a citation that appears at the end of the sentence.

Incorrect	`The presence of hairs on a leaf can increase light reflectance from the leaf surface. (Pearman 1966)`
Incorrect	`. . . leaf surface (Pearman 1966.)`
Correct	`. . . leaf surface (Pearman 1966).`

Work by Two Authors

Put the senior author's name first. (The senior author is the one whose name appears first after the title. It is generally assumed that he or she had major responsibility for writing the paper.)

```
In a study by Rutowski and Abrams (1963). . . .

Woods and Howell (1924) reported that. . . .
```

Work by Three or More Authors

Here, you may cite the senior author's name followed by the abbreviation *et al.* (from the Latin phrase *et alii,* meaning "and others"). Although some journals specify that *all* authors must be mentioned in the first citation, they allow the use of *et al.* in subsequent citations of the same reference. Some authors do not italicize (or underline) *et al.*

> White-lined bark beetles are attracted to the odor of rotting wood (Bateson et al. 1972).

> Dodson et al. (1977) have suggested that. . . .

Notice the use of *have* (not *has*) here. The subject (Dodson and others) is plural and therefore must have a plural verb.

Two or More Works by the Same Author

Give the author's name followed by the dates of the works in chronological order. Use commas to separate the dates.

> The circulatory system of this species has been described in detail by Banks (1978, 1980, 1983).

Two or More Works by the Same Author in the Same Year

Use letters to differentiate between references: *a* for the first work that you refer to, *b* for the second, and so on.

> Kannowski (1967a) conducted intensive field studies of this species and postulated (1967b,c; 1968) that males failed to defend territories under conditions of high population density.

Notice that you need not repeat the author's name if it already appears earlier in the sentence and you are not citing additional authors. Also note the presence of the semicolon, for clarity, between the letter, *c*, and the date, 1968, of the next work.

Two or More Works by Different Authors

To cite more than one reference within the same parentheses, list them in chronological order and separate them using commas. Some writers list a series of references alphabetically by the senior author's name.

> Several models have been proposed to account for this phenomenon (Wright 1935, Abrams and Chen 1960, Diaz 1980).

Work with No Author

To cite pamphlets, government documents, or other sources with no identified authors, refer to the author as "anonymous" (abbreviated, Anon.).

```
Creeping bellflower has been reported from both Mad-
ison and Oneida Counties of New York State (Anon.
1986).
```

A Specific Part of a Work

If you need to refer to a particular table, figure, or portion of a source, include this information within the parentheses.

```
(Chan 1955, Fig. 6)

Roberts (1986, chap. 1-3) discusses the evolutionary
basis of. . . .
```

A Direct Quotation

Some authors give the page number in parentheses, separated from the date by a comma. Others supply just the author and date.

```
Hassan (1984, p. 3) has written, ". . . .

Moore (1932) concluded, ". . . .
```

Unpublished Works

Occasionally you may need to cite a personal conversation with someone, an informal written communication, or the unpublished data of a colleague. Professional ethics dictate that you obtain the person's permission to include such material in your paper. For each citation, use the last name and initials of the person who supplied the information, followed by a word or two of explanation, either written out in full or abbreviated. Do not list these unpublished materials in the Literature Cited section because they are inaccessible to your readers.

```
T. Franklin (personal communication) has suggested
that. . . .

This species has also been found along the shores of
lakes and ponds (J. A. Fowler, unpublished).
```

If you refer to your own unpublished data from a *different* study, give your name followed by "unpublished" or "unpub."

Professional biologists are often familiar with their colleagues' manuscripts and may refer to these in their own writing. When an article or book has been accepted for publication but has not yet been printed, it is regarded as "in press." This status is indicated in the parentheses instead of the publication date.

```
Similar observations have been made by Kane (in
press).
```

If you cite a work in press, list it with already published works in the Literature Cited section. Anyone who wants to track it down will be able to do so by looking up recent issues of the journal in which it is scheduled to appear.

Works You Have Not Consulted Directly

Avoid referring to sources you have not read yourself; the reader assumes you have first-hand knowledge of all the works you have cited. Occasionally, however, you may need to cite an important source that is inaccessible to you. If so, specify where you acquired your second-hand knowledge of this source.

```
Needham (1954, cited in Vance, 1968)
```

List both sources in your Literature Cited section.

Modifications of the Name-and-Year Method

Some authors use a slightly different format that includes a comma between the name and year when both appear in parentheses. Thus, (Linsley 1966) would be written (Linsley, 1966). A semicolon is used to separate sources appearing together in parentheses: (Elliott, 1957; Lane, 1971).

Number Method

In this format, each source is given a number. The in-text citation consists of this number enclosed in parentheses. Full information about all sources cited is given in the Literature Cited section at the end of the paper.

```
Sap pressures in Douglas fir become more negative as
relative humidity rises (1).
```

If you mention a particular source more than once, use the same number originally assigned to it. To include two or more works within the same parentheses, list them in numerical order and separate them by commas.

```
Plantago major is common in heavily trampled areas,
such as the edges of roads and sidewalks (3, 8, 14).
```

Numbers are assigned to references in one of two ways, depending on journal specifications and the conventions followed in particular subdisciplines of biology. In one version, sources are numbered in the order in which they are cited in the text and listed in numerical order in the

Literature Cited section. In the second variation of the number method, sources are alphabetized, numbered in alphabetical order, and listed alphabetically in the Literature Cited section.

The number format for citing sources is less cumbersome and distracting than the name-and-year method, especially when you need to refer to many works one after another. However, it is much less informative for the reader. If the author or publication date of a particular work is important to your discussion, you must add this information to the sentence.

> Smith (11), studying three species of tree frogs in South Carolina, was the first to observe that. . . .

> This species was not listed in early floras of New York; however, in 1985 it was reported in a botanical survey of Chenango County (13).

As in the name-and-year format, if you need to refer to specific portions of a source, make this clear in the citation. Some authors also include the page number for a direct quotation.

> My data thus differ markedly from Markam's study on the same species in New York (5, Figs. 2 and 7).

> "Tailbeating" behavior, as defined by Cheevers and Briggs (3, p. 25), is ". . . .

■ Put citations where they make the most sense.

Whichever documentation system you use, put each citation close to the information you wish to acknowledge. Do not automatically put cited material at the end of every sentence. The following statement, for example, is ambiguous.

> Pollination of <u>Linaria vulgaris</u> has been studied in both the field and the laboratory (Arnold 1962, Howard 1979).

Did Arnold do his studies in the field and Howard in the laboratory? Or Howard in the field and Arnold in the laboratory? Or both authors in both settings? Moving the first citation clarifies the situation:

> Pollination of <u>Linaria vulgaris</u> has been studied in both the field (Arnold 1962) and the laboratory (Howard 1979).

If you actually mean to say that both Arnold and Howard did both types of studies, then you are better off rewording the sentence:

Arnold (1962) and Howard (1979) have studied the pollination of Linaria vulgaris in both the field and the laboratory.

■ Do not cite sources for information regarded as common knowledge in a particular field.

It is not strictly wrong to do this, but it is unnecessary. If certain material is well known and fundamental to a particular field, it is not necessary to cite sources. For example, you need not cite your textbook or other references to say that living organisms are composed of cells or that meiosis in higher animals gives rise to haploid gametes. Such material is general knowledge that is familiar to the audience of any biological paper. Similarly, any subdiscipline of biology has information that is regarded as elementary and basic by anyone working in the field.

How do you decide what is common knowledge and what is not? This ability comes with experience as you become more familiar with a subject and the literature on it. For academic assignments, the background knowledge you share with your classmates will be your best guide. If in doubt, ask someone more experienced than yourself, or cite the source anyway.

■ Use citations carefully.

Citations allow you to acknowledge the work or ideas of others; they also inform the reader. Do not pack your text with citations simply to demonstrate that you've done your homework and are intimately familiar with the literature.

Many studies have been made of the factors influencing variable mate guarding in the dragonfly Plathemis lydia (Brubaker and Nakhimovsky 1962, Darby 1963, Kraly and Kraly 1964, Domack and Joy 1965, Angell 1974, Napolin 1975, Hayes 1978, Pegg 1980). . . .

Vague reference to a huge number of works does little to inform the reader or further the progress of a focused discussion. Decide which of these sources are *most* important and why. Refer to them in a meaningful way so that they contribute to your discussion:

Many studies have been made of the factors influenc-

```
ing variable mate guarding in the dragonfly Plath-
emis lydia. Brubaker and Nakhimovsky (1962) found
that. . . , whereas Darby (1963) showed that. . . .
Experimental manipulations of male density at breed-
ing sites by Angell (1974) and Hayes (1978) sug-
gested that. . . .
```

Watch out for unintentional sexism when you refer to authors. Do not automatically assume that every biologist is male.

```
Brown (1985) investigated bacterial resistance to
the antibiotic chloramphenicol. She suggested
that. . . .
```

PREPARING THE LITERATURE CITED SECTION

■ Understand the difference between a Literature Cited section and a bibliography.

A bibliography contains all the sources mentioned in the text, along with additional references on the topic. The Literature Cited section of a biological paper contains *only* the literature (sources) that have been *cited* (referred to) in the text. Even if you have acquired useful background knowledge by reading five articles and three books, do not list any of these in the Literature Cited section unless you have specifically mentioned them in the text. Bibliographies are not generally part of scientific papers.

The Literature Cited section is sometimes entitled References Cited, or simply References.

■ Report sources completely and accurately.

After typing up the main body of the paper, you may be tempted to race through the listing of references, presuming that no one looks at this part anyway. Do not underestimate the importance of a meticulously prepared Literature Cited section. Readers do look at this section. In their minds, the amount of attention you have given it reflects the care given to the rest of the paper. If your sources are reported sloppily, people may doubt your authority, integrity, and thoroughness as a researcher and writer.

The Literature Cited section also serves as an important source of references for readers who want further information on the topic. You owe these people accuracy and completeness. Even many published papers contain mistakes or missing information in the Literature Cited section. Few things are more frustrating for a researcher than to be told the wrong page numbers or journal volume for an article he or she needs to track down.

Long *before* you sit down to prepare the Literature Cited section, become familiar with the kinds of bibliographic details you will need for each type of source (article, book, paper in an edited collection of papers, and so on). As you read and take notes in the library, record all this information in a master list or on file cards (see p. 43). Doing so will save you time and help you avoid chaos later.

■ Use a conventional format for reporting your sources.

Biological journals have adopted various formats for the Literature Cited section of papers. Prospective authors prepare this section by carefully following the guidelines prescribed by the journal for which they are writing. Your instructor may have certain preferences; otherwise, follow the format used by recent papers on your topic, or use the format below.

The following rules illustrate the style used by the journal *American Naturalist*.

Journal Article with Single Author

Grant, P. R. 1981. Speciation and the adaptive radiation of Darwin's finches. Am. Sci. 69:653-663.

Put the author's last name first, followed by his or her initials. Give the publication date next, then the title of the paper. Only the first word of the title is capitalized (in addition to proper nouns, such as *Darwin* in this example). The title of the journal is capitalized but not italicized. Note that the title is abbreviated to save space and printing costs. To find out the conventional abbreviation for a journal, look at a copy of the journal itself or consult *BIOSIS List of Serials*, published each January by *Biological Abstracts*, or the *American National Standard for Abbreviations of Titles or Periodicals*. Abbreviations for medical journals are listed annually in the January issue of *Index Medicus*. If the title has no conventional abbreviation or if it is a single word, write it out in full.

After the journal title, give the volume number followed by a colon and the pages on which the article appears. For journals that number each issue separately (not continuously through a single volume), give the issue number in parentheses after the volume number.

Wolfram, S. 1984. Computer software in science and mathematics. Sci. Am. 251(3):188-203.

Journal Article with Multiple Authors

Via, S., and R. Lande. 1985. Genotype-environment interaction. Evolution 3:505-522.

The first or senior author is listed first, followed by one or more coauthors in the order in which they appear on the title page. Only the first author's names are reversed; the other authors appear with the initials first, followed by the last name. Use a comma to separate each author's name from the others.

Book

Cavalli-Sforza, L. L., and W. F. Bodmer. 1971. The genetics of human populations. Freeman, San Francisco.

Falconer, D. S. 1981. Introduction to quantitative genetics. 2nd ed. Longman, London.

The rules for authors and titles of books are similar to those for journal articles. After the title give the edition number for all editions after the first. Use a comma to separate the publisher from the place of publication.

Pamphlet or Government Document

Weins, J. A. 1975. Avian communities, energetics, and functions in coniferous forest habitats. U.S. For. Serv. Tech. Rep. WO-1:226-265.

Article or Chapter in an Edited Volume

Halliday, T. R. 1978. Sexual selection and mate choice. Pages 180-213 in J. R. Krebs and N. B. Davies, eds. Behavioural ecology: an evolutionary approach. Blackwell, Oxford.

List the article or chapter by the author's last name, followed by the date, title, page numbers, and full bibliographic information on the source. Note the order of information given here, the use of italics (underlining) for *in,* and the abbreviation for "editors" (eds). Also notice the British spelling of *behaviour* as opposed to *behavior.* Retain the original spelling for all references you report.

UNPUBLISHED MATERIAL

If the work has been accepted for publication and is in press, use this format:

Smith, P. P. 1987. Biology of yeasts. Howard and Davidson, New York (in press).

If the work is a thesis or dissertation, follow this style:

Brown, J. H. 1969. The life history and ecology of the northern lake chub (Couesius plumbeus) in the La Ronge region of Saskatchewan. M.Sc. Thesis, University of Saskatchewan, Saskatoon, Sask., Canada.

Note that the genus and species name of this fish are underlined (italicized), and only the genus is capitalized. See pages 113–115 for how to treat scientific and common names of organisms.

For other unpublished material (an informal communication with an author, a friend's raw data, a conversation with someone), cite the author and the nature of the material in the text, but do not list these sources in the Literature Cited section because they are not accessible to other readers.

Arrangement of References

American Naturalist uses the name-and-year method for literature citations. Therefore, references in the Literature Cited section appear alphabetically according to the senior author's last name. If a particular author or group of authors has published more than one work, list these in chronological order according to the publication date. If you have two or more works by the *same* author(s) with the *same* publication date, use letters to distinguish them. For example, two papers by Waldman in 1977 would be listed as 1977*a* and 1977*b*, with *a* assigned to whichever paper was first cited in the text.

Works by two or more different authors with the same last name are arranged alphabetically according to the authors' first initials. For example, an article by K. L. Dodge would come before one by M. A. Dodge.

Other Formats

There are many variations of the format given above. Many journals follow the conventions used in the *CBE Style Manual* (5th ed., 1983), published by the Council of Biology Editors. These are generally similar to *American Naturalist*'s guidelines except in certain details of punctuation and ordering of information. (The publication date, for example, appears last in the CBE format.) Other journals are distinguished by their own idiosyncratic styles: some specify that journal titles be italicized; some do not allow abbreviations; some omit the titles of journal articles; some call for certain information to appear in boldface type; and so on. It is easy to feel overwhelmed by the diversity of reference formats. However, the most important thing to remember is to be consistent. Choose one format and follow its rules carefully for every reference you list.

CHAPTER 5

Drafting and Revising

Expect to do several drafts of your paper before you are satisfied with the final product. Good writing is generally the product of careful *rewriting* or *revising* in which you evaluate your early attempts at organizing and expressing your ideas. In the process you end up scrutinizing the ideas themselves, as well as your own mastery of the subject. You may find that until you can express a concept clearly enough that others can understand it, you have not fully understood it yourself. Working through successive drafts also teases out *new* ideas, perhaps connections between disparate sets of data or new insights about the significance of your research. As you grope for more appropriate wording and a clearer structure, you give form and substance to thoughts still lurking in your subconscious. Gradually and sometimes painfully, you discover what it is you really want to say.

THE FIRST DRAFT

Many people assign too much importance to a first draft. Think of it as just a rough version of the paper, an exercise in organizing your thoughts and getting down the main ideas. Focus on content, not on prose style or mechanics. Because much of what you write now may not end up in the final version, it is unproductive to tinker with commas, spelling, or sentence construction. First get down the basic information; in later drafts you can look more closely at your writing style and such matters as grammar and punctuation.

How do you shift from doing research or taking notes to composing the first draft of the paper? Experienced writers use a variety of methods. The following ones may help you get started.

■ Devise a working title.

The title should reflect your most important findings and your purpose in writing the paper. What is the main point you wish to make? What new information or perspective does your study contribute? Deciding on a preliminary title early in the writing process helps you focus your thoughts and start drafting the paper. If necessary, you can revise the title later as your ideas take shape.

■ Make an outline or a rough plan.

All writers need some sense of how the various points they want to make are logically connected to one another. However, people vary tremendously with respect to how organized their thoughts must be before they start to write. Some writers find that a detailed, point-by-point outline is absolutely essential before they start the first draft. Others prefer to work out a more general framework that leaves them considerable flexibility. Some writers find any kind of outline confining. Instead, they plunge directly from their notes into the first draft, doing most of the organizing in their heads as they go along.

Systematic planning may be more appropriate for some sections of the paper than for others. For example, the Materials and Methods section of a research paper is fairly straightforward and contains many details that must be organized in a logical manner. Once you plot it out carefully, it may practically write itself. The same may be true of the body of many review papers. By contrast, the complexities of a Discussion section may be difficult to pin down and categorize. In fact, you may not be sure of what you want to say until you start putting words to paper and have worked through one or two drafts.

It pays to experiment with different forms of planning. If you don't take to outlines, try jotting down a rough list of the major topics or issues you need to discuss and working from there. You may find it easier to visualize the relationships between ideas if you make a "flow chart" with lines, arrows, or brackets linking groups of words or phrases. If you do rely on detailed outlines, remember that these should not be inflexible schedules. They can be stretched or restructured to conform with your writing as it develops. An outline can be a valuable tool to get you started, but it should never be an end in itself. It should never keep you very long from that most important act of plunging into the paper.

■ Start the easiest writing first.

Don't feel you have to draft the Abstract first, then the Introduction and remaining sections of the paper in strict order. Begin with whatever you feel is easiest and most straightforward. For many biologists this is the Materials and Methods section, followed by Results and Discussion. The Introduction and Abstract are often easiest to write last, once you have a better idea of what the paper is all about. The point is that once you start drafting *any* section of the paper, you have broken the barrier between you and the material, and the writing will gain momentum.

■ Talk to others.

Brainstorming with other people about the project may help get you started by giving you an audience on which to test out your ideas. Professional scientists rely heavily on such feedback from their colleagues, both in the more formal settings provided by conferences and on a day-to-day basis during lunch or coffee breaks. Intellectual exchange with others helps scientists keep their work on track and instills it with fresh insights. If you read the Acknowledgments portion of several scientific papers, you will see how important such discussions are to research and writing processes.

PRACTICAL SUGGESTIONS FOR REVISING

Revising will be most efficient if you break down the task into several successive steps. First consider the paper as a whole, checking the overall content, organization, coherence, and consistency of your argument. Try to resist the impulse to make small-scale changes — for example, changes in wording or punctuation — until you are satisfied with the main substance and structure of your paper. Then work on polishing your prose style, examining individual paragraphs and sentences closely for clarity, accuracy, and conciseness. Look for wordiness, awkward sentence constructions, improper word usage, faulty subject-verb agreement, excessive use of the passive voice, and other problems that may obscure your meaning and impede the flow of your writing. Finally, proofread the text for punctuation, spelling, and typographical errors, and inspect the final form of the manuscript (see Chapter 6).

In practice, of course, making revisions is not quite so orderly and mechanical a process. Also, "writing" and "revising" a paper are not totally separate activities: writers often make major and minor revisions *as* they are composing, not just after they have written a draft. Nevertheless, your writing may be more productive if you set some priorities for yourself, focusing on big changes first, small changes later. Becoming more aware of how you write will help you develop the drafting and revising strategies that work best for you.

Do not discard your early drafts; keep everything at least until you have submitted the final version of the paper. Sometimes parts of a first or second draft end up sounding better than the corresponding sections of later ones. Put both the page number and the number of the draft on each piece of writing in case you mix up parts of different drafts. You may wish to photocopy each draft in case you lose the original.

If you type your drafts or write them out in longhand, the "cut-and-paste" method saves time and effort when you need to move chunks of text from one place to another. Use scissors and tape or glue to patch the new version together, rather than typing or writing everything over again.

Allow plenty of time for revising your paper. Set each draft aside for several hours or several days; when you return to your writing you'll view it more objectively and work more productively. Also ask one or two of your friends to criticize the manuscript. Pick people who will give honest and thoughtful opinions and not just say what they think you want to hear. Professional biologists rely heavily on criticism from their colleagues as they prepare manuscripts. Once papers are submitted to journals they are subjected to further scrutiny by editors and outside reviewers. This review process, as formidable as it may sound, gives authors invaluable feedback about their work and helps to ensure high standards for research and writing.

REVISING WITH A WORD PROCESSOR

Because word processors have so simplified the process of producing clean copies of text, they are extremely helpful for revising and proofreading. Using a few keystrokes, you can add, delete, or change letters or words; move whole sections of text from one site to another; and make sweeping changes in the format of the whole manuscript. Seeing your writing appear neatly on the screen in front of you may make it easier to shift from the role of writer to critical reader and back again, as you work through the current draft of the paper.

Here are some suggestions about using the computer most productively.

1. Because it is so simple to alter the surface features of the text, you may be tempted to make even the early drafts "perfect," with every comma in place and every typographical error corrected. However, doing so will distract you from the more important job of articulating and organizing your main ideas. Wait until you are satisfied with the paper as a whole before you spend time and effort on local changes in the text.

2. Do not confine all your revising to the computer screen. The computer shows you only a small part of the manuscript at one time, but some

problems may be more noticeable when you can spread out several pages of the paper in front of you. For this reason, it makes sense to get frequent printouts and make additional changes by hand directly on the manuscript. Keep all the printouts even after you have incorporated the changes in your file. You may still need to refer to an early draft before you are finished with the paper.

3. Many writers who have become addicted to word processing find that a blank screen and the familiar hum of the computer are just the stimuli they need to start writing. Other people work better if they do their early composing in longhand and save the last stages of revising and proofreading for the word processor.

4. You may wish to use a style-analysis program to help you revise your writing. Such programs call attention to potential problems in the paper, such as too many short sentences or overuse of *to be* verbs, as well as missing parentheses or quotation marks. Of course, they cannot meaningfully evaluate your prose; you must still make those judgments yourself.

5. Spelling checkers will locate misspelled words or typos such as *reserch* instead of *research,* or *studyy* instead of *study*. However, they cannot detect typographical errors that are legitimate words, such as *samples* instead of *sample*. They also can't distinguish between the various spellings of words such as *two, to,* and *too;* or *principle* and *principal*. Finally, although spelling checkers have sizeable dictionaries, these may not include many of the words you are using, such as specialized terms and proper nouns. However, many spelling checkers allow you to enter such words into their dictionaries.

6. The "search" or "find" feature of word processing programs is invaluable for locating specific words or phrases in the text that you need to change. For example, suppose you realize that throughout the paper you have typed the genus *Gomphus* incorrectly as *Gomphis*. On command the computer can locate every place where *Gomphis* appears in the text so that you can correct each misspelling. This process is even easier if your program has an automatic find-and-replace function, in which case a single command will change every *Gomphis* to *Gomphus*. Similarly, if you suspect that you have not only misspelled *Gomphus* but also failed to underline it occasionally, you can conduct a second search for all instances of "Gomphis" and edit these accordingly.

In many word processing programs, to use the search feature you must be at a point in your text that *precedes* the material you need to locate. This is because these programs can only search forward, not backward. If you are not sure where this point is, then position the cursor at the very beginning of the manuscript. Other word processing programs can search both forward and backward. Also note that you must specify the word you wish to locate *exactly* as it appears in the paper; if it is underlined or capitalized or put in boldface type, you must include these features. Otherwise the computer cannot recognize it.

7. Finally, but most important: it is wise to make a duplicate, or backup, file for your paper on a second computer disk as a safeguard against inadvertently erasing the first file. Remember to update the backup file each time you make revisions. When you are not working on the computer, store the disks in different places.

CHECKING CONTENT AND STRUCTURE

■ Improve logic, continuity, and balance.

In a research paper you need to keep track of your argument. What major question are you addressing? What hypotheses have you attempted to test? Do the data you present in your Results directly relate to these hypotheses and to the conclusions put forth in the Discussion? A good scientific paper is somewhat like a mystery story with the "solution" well supported by a carefully crafted body of evidence in the Results section. However, the scientific writer has also gradually prepared the reader for the major conclusions, which should come as no sudden surprise.

Although a review paper has a less formal structure, it also needs to be built around a logical train of thought. Readers need to understand your rationale for choosing the topic. They need a sense of movement as you develop your main points, and they need to see the paper end on a satisfying note through summarizing statements and conclusions.

Once you have completed a rough draft, try to visualize the paper as a unit. Is its structure clear? Do you lead readers along your line of reasoning point by point, paragraph by paragraph, making clear transitions from one topic to the next? Or do they have to grope their way along, guessing the connections? Are any topics underdeveloped or overdeveloped, throwing the paper out of balance? Do you maintain a consistent style and approach? Compare your introduction with the rest of the paper. Have you delivered to the reader what you promised at the beginning? Are your initial statements compatible with your concluding ones?

Sometimes large-scale revisions can be accomplished relatively effortlessly by moving sentences or even whole paragraphs from one place to another. Usually, however, you need to do substantial rewriting — paring down overwritten passages, modifying highly speculative comments, developing poorly expressed ideas, adding new clarifying material, revising the beginnings and ends of paragraphs so that you can fit them into new contexts. Such revisions can be frustrating, even painful. However, they lead to a new product: a clearer, more organized version of the paper.

■ Omit unnecessary material.

Even when you feel you have "nothing to say," you may end up with more material than you anticipated, perhaps more than you really need. As you check the manuscript for logic and balance, be alert for places where you strayed off the topic. The Materials and Methods section may include procedures that do not pertain to any of the data discussed. The Results can easily become a grab bag of miscellaneous data, only some of which are important to the main story. The Discussion may be weighed down by predictions impossible to test in the near future or by rambling comments of only peripheral importance. Check that *every* point is relevant to your main objectives. Be ruthless. Cut any extraneous material — text, table, or figure — no matter how hard you worked on it.

■ Check for completeness and consistency.

Writing a research or a review paper requires you to keep track of many details. It is surprisingly easy for the manuscript to develop inconsistencies or to be missing needed bits of information. Here is a checklist of questions to ask yourself:

Is the title specific and informative?
Does the Abstract and/or Table of Contents include all relevant parts of the paper?
Does the Introduction state your main objectives, hypothesis, or thesis?
Have you included enough information in the Materials and Methods section to enable someone else to repeat your study?
Have you explained in Materials and Methods the procedures for collecting all the data presented in the Results?
Are figures and tables numbered consecutively in separate series?
Is every figure and table cited correctly in the text?
Do the data in each figure or table agree with your in-text discussion?
Do any figures or tables present conflicting data?
Are data in related figures or tables shown in a consistent manner?
Is each table and figure understandable apart from the text?
Are any important Results missing?
Have you used enough subheadings to guide the reader?
Does the Discussion address the major implications of your findings?
Have you considered problems, inconsistent results, and counterevidence?
Have you cited all necessary sources?
Are all sources cited in the text listed in the Literature Cited section?
Does the Literature Cited section include any sources not cited in the text?

IMPROVING PARAGRAPHS

■ Present coherent units of thought.

Paragraphs are not just chunks of text; at their best, they are logically constructed passages organized around a central idea often expressed in a *topic sentence.* A writer constructs, orders, and connects paragraphs as a means of guiding the reader from one topic to the next, along a logical train of thought.

Topic sentences often occur at the beginning of a paragraph, followed by material that develops, illustrates, or supports the main point.

> *The teeth of carnivorous and herbivorous vertebrates are specialized for different ways of life.* Those of carnivores are adapted for capturing and subduing prey and for feeding largely on meat. Dogs and cats, for example, have long, sharp canines used for piercing, and molars and premolars equipped for cutting and tearing. By contrast, herbivores such as cows and horses have teeth specialized for feeding on tough plant material and breaking down the indigestible cellulose in plant cell walls. Their molars and premolars have large, ridged surfaces useful for chewing, gnawing, and grinding.

Do not distract the reader by cluttering up paragraphs with *irrelevant* information. For example, in the passage below, the second sentence does not relate to the paragraph's main point as stated in the topic sentence preceding it. We can strengthen the paragraph by deleting the second sentence altogether.

> *Tanner (1981) sheds new light on the processes that may have been critical in the evolution of early hominids from chimpanzee-like ancestors.* Tanner also includes much interesting information about chimpanzee social life. She suggests that a critical innovation might have been the extensive use of tools (initially, organic materials such as bones or sticks) by females for gathering. Such tools, she argues, would have been employed long before weapons were being manufactured by men for hunting large game.

How long should a paragraph be? There is no set rule; paragraph length depends on the writer's topic, coverage, format, and purpose. As a working rule, aim for four to six sentences, then shorten or lengthen the passage as needed. Avoid one-sentence paragraphs. Used skillfully, for example in informal essays or in fiction, they give variety and emphasis. However, they may express a poorly developed idea or one that really belongs in a neighboring paragraph. Conversely, very long paragraphs can get unwieldy and confusing. Usually such passages contain more than one idea and can be divided.

Keep in mind that it is not enough to group related sentences together into a paragraph. You must also *demonstrate* these relationships; otherwise, the reader cannot follow your line of thought. Look at the following passage from a lab report:

> Plants provide a constant supply of oxygen to our atmosphere. Both plants and animals depend on oxygen in the utilization of their food. During the process of photosynthesis, plants consume carbon dioxide and release oxygen.

The reader has to struggle to find the point of this paragraph. One problem is that the sentences seem to be thrown together haphazardly. Another is that the writer provides no conceptual links from one sentence to the next. Here is a revised version of the paragraph:

> During the process of photosynthesis, plants consume carbon dioxide and release oxygen. Both plants and animals depend on oxygen in the utilization of their food. Thus, plants provide a constant supply of this needed substance to our atmosphere.

In the revision the sentences are arranged more logically. Moreover, transitional elements — *thus* and *this needed substance* (referring to oxygen) — help link the last two sentences and clarify the main point of the paragraph.

Here is another paragraph to illustrate the importance of showing the reader a clear pathway of thought. Notice again the use of transitional words, along with the repetition of selected elements, to clarify the relationship between sentences:

> According to sociobiological theory, the production of individual ova is costly relative to the production of sperm. *Therefore,* ova (or females) are the limiting factor in male reproductive success, and natural selection will favor those males who can compete effectively with other males for fertilizations. We can predict that males *will vary* greatly in fitness depending on their competitive abilities. *By contrast,* females *will vary* in their abilities to convert available resources into gametes and ultimately into viable offspring.

■ Make paragraphs work as integrated parts of the text.

Apart from the Abstract, most paragraphs are not isolated entities. Instead they must mesh together smoothly as structural components of each section of your paper. This means that the beginning of one paragraph must "fit" with the end of the previous paragraph — you need to bridge the gap between these two passages gracefully. If you jump abruptly from one topic to the next, the text will seem choppy and disorganized and the reader will be confused. This problem often crops up in Results and Materials and Methods sections when you are presenting many different kinds of information.

You can use the same techniques for linking sentences *within* paragraphs to making transitions *between* paragraphs. These include the repetition of key words or ideas and the use of transitional "markers" (*furthermore, for example, a second point, by contrast, on the other hand,* and so on) to signal to the reader that you are either developing the same idea or moving on to a different one.

■ Vary your sentences.

Pay attention to the structure, length, and rhythm of your sentences. People "hear" writing as they read; if your prose is unvarying and one-dimensional you will not get your message across as effectively.

The following paragraph is dominated by short, choppy sentences:

> Many doves showed "nest-calling" behavior. They assumed a position with the tail and body axis pointing slightly upwards. In this posture they flicked their wings. This behavior was seen in both sexes. It was especially common in males. I saw it performed both on and off the nest.

We can make this passage more readable (and therefore more interesting) by combining related sentences. This eliminates the distracting, choppy style and makes the paragraph more effective as a unit:

> Many doves showed "nest-calling" behavior in which they assumed a position with the tail and body axis pointing slightly upwards and flicked their wings. This behavior, especially common in males, was performed both on and off the nest.

WRITING CLEAR, ACCURATE SENTENCES

■ Use words that say precisely what you mean.

Do not give in to the temptation to use a word that "sounds right" unless you are absolutely sure it is appropriate. Sentences that are otherwise perfectly effective can be ruined by a single word or phrase that is wrong for the context. Here are some examples from student papers for a mycology class:

INCORRECT	Kohlmeyer (1975) sees the pit plug of red algae as being *heavily* related to the plug of Ascomycetes.
CORRECT	Kohlmeyer (1975) sees . . . *closely* related. . . .

INCORRECT	Many mycologists have spent years *researching* for data suggesting a red algal ancestry with higher fungi.
CORRECT	Many mycologists have spent years *searching*. . . .
INCORRECT	The evolutionary origin of the higher fungi has *harassed* scientists for many years.
CORRECT	The evolutionary origin . . . has *puzzled* scientists. . . .

Buy a dictionary and consult it often. Before you use a technical term make sure you understand its meaning. If in doubt, look it up in the glossary of an introductory text. The reference section of your library also has specialized dictionaries (see p. 132) and scientific encyclopedias.

The following words are often used incorrectly in biological papers:

Affect: (as a verb) to influence or to produce an *effect*.
Effect: (as a noun) result; (as a verb) to bring about.

Nutrient concentration was the most important factor *affecting* population size.

Marking each ant on its thorax with enamel paint produced no apparent *effect* on its behavior.

We hope that further studies of these endangered species will *effect* a major change in the allocation of funds by the federal government.

Comprise: to contain. Do not use *comprise* when you mean *constitute*, or make up.

The vertebrate central nervous system *comprises* the brain and the spinal cord.

The brain and the spinal cord *constitute* the vertebrate central nervous system.

Correlated: (see p. 24).
Interspecific: between or among two or more different species.
Intraspecific: within a single species.

Leone (1960), studying four species of sandpipers at a Minnesota lake, found marked *interspecific* differences in food preferences.

This plant shows little *intraspecific* variation in flower coloration; generally, the petals are pale yellow with a distinct orange spot at each tip.

Its/it's: Its is the possessive form of *it*. Do not confuse it with *it's*, a contraction of *it is* or *it has*.

Each calf recognized *its* own mother.

It's not clear whether rainfall or temperature is the more important factor.

Note, however, that contractions are generally avoided in formal biological writing.

Principal: (as an adjective) most important. Do not confuse it with *principle,* a noun, meaning a basic rule or truth.

> The *principal* finding of this study was that excessive drinking by rats did not cause significant increases in blood pressure.

> According to Heisenberg's Uncertainty *Principle,* we can never simultaneously determine the position and the momentum of a subatomic particle.

Random: (see p. 24).

Significant: (see pp. 23–24).

That/Which: Use *that* to introduce *restrictive* or defining elements — phrases or clauses that limit your meaning in some way.

> The rats *that had been fed a high calorie diet* were all dead by the end of the month.

Here the italicized portion restricts the meaning of *rats;* we are referring to *only* those specific rats.

Use *which* to introduce *nonrestrictive* or nondefining elements — word groups that do not limit your meaning but rather add additional information. Because this information is not vital to the integrity of the sentence, you can omit it without substantially changing the original meaning. The following sentence uses *which* instead of *that;* here the writer is speaking more generally, not calling attention to a *particular* group of rats.

> The rats, *which* were fed a high calorie diet, were all dead by the end of the month.

Misuse of *that* or *which* may make a sentence confusing:

> Plants, which grow along heavily traveled pathways, show many adaptations to trampling.

Here it sounds as if *all* plants grow along heavily traveled pathways. Actually the writer is referring only to a particular group of plants. The nonrestrictive clause needs to be replaced by a restrictive one limiting the meaning of *plants:*

> Plants that grow along heavily traveled pathways show many adaptations to trampling.

Use commas to set off nonrestrictive elements introduced by *which* — a *pair* of them if the element appears in the middle of the sentence.

> These data, which are consistent with those of other researchers, suggest several questions about the significance of wing positions in the thermoregulation of Arctic butterflies.

Do *not* use commas with restrictive elements and *that.*

Unique: one of a kind. A thing cannot be *most, very,* or *quite* unique; it is simply unique:

INCORRECT	Females of this species have a *very unique* horny projection from the dorsal part of the thorax.
CORRECT	Females of this species have a *unique.* . . .

■ Avoid slang.

Slang is the informal vocabulary of a particular group of people. Some slang words eventually make their way into standard English; however, most soon become outdated and are replaced by others. Slang usually has no place in scientific writing even if you put in quotation marks.

SLANG	The controversy over the evolutionary origin of the Ascomycetes and Basidiomycetes dates back almost a hundred years, but it has only recently moved to the cutting edge of interest.
STANDARD	Recently, scientists have become more interested in the evolutionary origin of the Ascomycetes and Basidiomycetes, a topic that has been controversial for almost a hundred years.
SLANG	Barr's (1980) statement is merely a "cop out" because he refuses to acknowledge that there are major morphological differences between the two groups.
STANDARD	Barr (1980) fails to address this issue because he refuses to acknowledge that. . . .

■ Revise misplaced modifiers.

Modifying words, phrases, or clauses should relate clearly to the words they are meant to describe. If they do not, the sentence may be confusing, even ludicrous.

FAULTY	After marking its hindwings with enamel paint, each damselfly was released within 1 m of the capture site.
REVISED	After marking its hindwings with enamel paint, I released each damselfly within 1 m of the capture site.

Unless these insects were particularly clever at marking themselves, we need to make the intended subject of this sentence (the writer) more evident. Using the active instead of the passive voice helps here. (See also pp. 110–112.)

FAULTY After mating, the sperm are stored in a sac within the female damselfly's body.

REVISED After mating, the female damselfly stores sperm in a sac within her body.

Who is actually mating, the sperm or the dragonfly? The revised version corrects the ambiguity.

FAULTY This behavior has only been reported in one other primate.

REVISED This behavior has been reported in only one other primate.

Modifiers such as *only, even, almost,* and *nearly* should be placed next to the most appropriate word in your sentence, depending on your meaning. In the first sentence, the position of *only* before the verb suggests that behavior may have been observed in more than one primate but *reported* in just a single species. If you really want to say that this behavior has been observed *and* reported in just this one species, then *only* must modify *one* as in the revised version. In the following sentence, *only* does limit the meaning of the verb in accordance with the writer's meaning.

This behavior has only been observed casually, not reported formally.

■ Avoid vague use of *this*, *that*, *it*, and *which*.

Do not use these pronouns on their own to refer to whole ideas; the reader often gets lost. To avoid confusion, use clear, specific wording.

VAGUE We could not predict the number of adult males likely to visit each breeding site because male density in the surrounding forest varied greatly from day to day. This is typical of most field studies on this species.

What is "typical of most field studies"? Varying male densities, the difficulty of being unable to predict male numbers at the breeding site, or both? The revision eliminates this ambiguity:

SPECIFIC We could not predict the number of adult males . . . from day to day. Varying male density is typical of most field studies. . . .

■ Make comparisons complete.

Add words if necessary to make comparisons or contrasts accurate and unambiguous.

AMBIGUOUS Average body length in *Libellula pulchella* is longer than *Plathemis lydia*.

Here, you want to compare the body length of one species with the *body length* of another, as the revision clearly does:

UNAMBIGUOUS Average body length in *Libellula pulchella* is longer than that in *Plathemis lydia*.

AMBIGUOUS Bullfrogs were more abundant than any amphibian at Site A.

Because bullfrogs themselves are amphibians, you need to revise the wording to avoid confusion:

UNAMBIGUOUS Bullfrogs were more abundant than any other amphibian at Site A.

■ Make each verb agree with its subject.

Do not lose sight of the subject in a sentence by focusing on modifying words, such as prepositional phrases, occurring *between* the subject and the verb.

The *size* of all territories *was* [not *were*] reduced at high population densities.

The *zygote* of the Ascomycetes *develops* [not *develop*] into ascospores.

Both sentences above have singular subjects and therefore need singular verbs.

Compound subjects are subjects with two or more separate parts that share the same verb. When these parts are connected by *and*, they need to be matched with plural verb forms.

The *color and shape* of the beak are [not *is an*] important taxonomic *features* [not *feature*].

When the parts of a compound subject are linked by *or* or *nor*, make the verb agree with the part that is closest to it. If one part of the subject is singular and the other plural, then put the plural part second and use a plural verb.

Under experimental conditions, *neither* the newly hatched offspring *nor* its older sibling *was tended* by the parents.

Here the parts of the compound subject are *offspring* and *sibling;* these are joined by *nor*. If the second part had been *siblings*, you would have needed a plural verb (*were tended*).

Do not confuse compound subjects with singular subjects that are linked to other nouns by prepositional phrases (such as *in addition to, along with, as well as*).

The dominant *male,* along with his subordinates, *protects* [not *protect*] the offspring when the troop is threatened by predators.

When you refer to a particular quantity of something as a single unit, treat it as singular.

Before each experiment, *10 ml* of distilled water *was* added to each vial.

Many scientific terms have a Latin heritage and are used in both the singular and plural form, depending on the context. Make sure the verb agrees with the subject.

The *larva* of the monarch butterfly *feeds* [not *feed*] on milkweed.

Here are the singular and plural forms of some words commonly used in biology.

SINGULAR	PLURAL
alga	algae
analysis	analyses
bacterium	bacteria
criterion	criteria
datum (*rarely used*)	data
flagellum	flagella
fungus	fungi
genus	genera
hypothesis	hypotheses
inoculum	inocula
larva	larvae
matrix	matrices
medium	media, mediums
nucleus	nuclei
ovum	ova
phenomenon	phenomena, phenomenons
phylum	phyla
pupa	pupae
serum	sera, serums
spectrum	spectra
stimulus	stimuli
stratum	strata
taxon	taxa
testis	testes
villus	villi

The word *species* can be either singular or plural: one species of cat, three species of toads.

■ Put related elements in parallel form.

When you link two or more words, phrases, or clauses in a sentence, put them in the same grammatical form. Such parallelism makes your writing easier to read and emphasizes the relationship between items or ideas.

FAULTY	These two species differ in color, wingspan, and where they typically occur.
PARALLEL	These two species differ in color, wingspan, and habitat.
FAULTY	Both populations of plants had high mortality rates at sites that were dry, windy, and where there were frequent disturbances.
PARALLEL	Both populations of plants had high mortality rates at sites that were dry, windy, and frequently disturbed.
FAULTY	Male bluntnose minnows promote the survival of their offspring by agitation of the water over the eggs and keeping the nest free from sediment.
PARALLEL	Male bluntnose minnows promote the survival of their offspring by agitating the water over the eggs and keeping the nest free from sediment.

In the first example, the clause, *where they typically occur,* has been replaced by a single noun, *habitat,* for parallelism with the two preceding nouns (*color, wingspan*). Similarly, in the second example, parallelism has been achieved by putting *where there were frequent disturbances* into adjective form, *disturbed,* to agree with *dry* and *windy*. In the third example, the noun *agitation,* has been changed to a present participle (verb form ending in *-ing*), making two parallel phrases, "*agitating* . . ." and "*keeping* . . ."

■ Write in a direct, straightforward manner; avoid jargon.

Scientific writing has the reputation of being dry, monotonous, and hard to understand. Consider the following passage from a published paper.

> The data of this study suggest that such a handling procedure not only effects a diminution of emotional behaviour as indexed by decreases in the duration of the immobility reaction, but also as indexed by other measures of fear (freezing and mobility in the

open field). To the extent that the handled and non-handled groups differed in the predicted direction with respect to these indices of fear, distress vocalizations in the open field were shown to be significantly less frequent in the relatively more fearful (non-handled) group.

(Ginsberg, Braud, and Taylor 1974, p. 748)

A paper may contain exciting results or brilliant insights, but still be tedious to read because of the author's unengaging prose. Whether it is a manuscript submitted for publication or a laboratory report for a biology class, an impenetrable style may keep it from getting the attention it deserves.

Frequently, the problem is the use of *jargon* instead of simpler, more straightforward writing. Broadly, jargon is the technical language of some specialized group, such as biologists. More specifically (and negatively), jargon is long-winded, confusing, and obscure language. Writers of jargon use esoteric terms unfamiliar to most of their readers. They rely heavily on long sentences, big words, a pompous tone, and ponderous constructions in the passive voice.

Jargon One hour prior to the initiation of the experiment, each avian subject was transported by the experimenter to the observation cage. The subject was presented with various edible materials, and ingestion preferences were investigated utilizing the method developed by Wilbur (1965). When data collection was finalized, the subject was transferred back to the holding cage.

Revised One hour before the experiment, I put each bird in the observation cage, where feeding preferences were studied using Wilbur's (1965) method. The bird was then returned to the holding cage.

The revised passage conveys the same information as the jargon-ridden one, but more simply, directly, and concisely.

The following passage from a successful textbook (Raven and Johnson 1986, p. 331) shows that straightforward writing allows the reader to grasp difficult concepts with a minimum of energy. Notice that *similes* ("like so many Mexican jumping beans") and *metaphors* ("passengers in a car with no driver") have their place even in scientific writing when they are used skillfully to illustrate a point.

Transposition is a form of gene transfer that occurs both in bacteria and in eukaryotes. Transposing genes do not stay put on chromosomes. Every once in a while, after many generations in one location, a transposing gene will abruptly move to a new position on the chromosomes, the location of its new residence apparently chosen at random. Transposing genes move about the chromosomes like so many Mexican jumping beans. These nomadic genes behave in this unusual way because they are carried along as parts of randomly moving genetic elements called transposons, passengers in a car with no driver.

Do not assume that to sound like a biologist you must write dry, stilted, and boring prose; or that complex ideas must be couched in equally complex, convoluted sentences. Biology instructors, along with editors and readers of biological journals, prefer clear, straightforward writing — simple but effective prose that quickly illuminates the author's results and ideas. Such writing has its own kind of grace and elegance.

AVOIDING WORDINESS

First drafts are usually labored and wordy because you have been focusing on just getting down your ideas. As you revise, examine each sentence carefully. Could you say the same thing more succinctly without jeopardizing the content? Lifting just one excess word from a sentence can liven its rhythm and intensify its meaning, making your prose carry more weight.

■ Omit unneeded words; shorten wordy phrases.

WORDY	On two occasions, I succeeded in observing a mating pair for the entire duration of copulation.
CONCISE	I observed two pairs for the duration of copulation.
WORDY	There now is a method, which was developed by Jones (1973), for analyzing the growth of rotifer populations.
CONCISE	Jones (1973) developed a method to analyze the growth of rotifer populations.
WORDY	The sample size was not quite sufficiently large enough.
CONCISE	The sample size was not large enough.
MORE CONCISE	The sample size was too small.
WORDY	The root cap serves to protect the cells of the root meristem as the root is growing through the soil.
CONCISE	The root cap protects the cells of the root meristem as the root grows through the soil.

WORDY	The eggs were blue in color, and they were covered with a large number of black spots.
CONCISE	The eggs were blue with many black spots.

Common modifiers such as *very, quite,* and *rather* can often be cut from sentences. You may use such words routinely without asking yourself if they are essential to your meaning.

FAULTY	The data in Table 1 are *very consistent* with Leshchva's (1966) model.
REVISED	The data in Table 1 are *consistent* with. . . .
FAULTY	Males guarding eggs are *quite aggressive* toward juveniles and females.
REVISED	Males guarding eggs are *aggressive* toward. . . .

In summary, biological writing is plagued by countless wordy phrases, often placed at the beginning of a sentence, and by "empty" words and phrases that add little to the author's meaning. Some of these are listed below. (See also Day 1983.) You will probably think of many others to add to the list.

WORDY	**CONCISE**
a second point is that	second, secondly
more often than not	usually
it is apparent that	apparently
at the present time	now
in previous years	previously
owing to the fact that	because
because of the fact that	because
in light of the fact that	because
it may be that	perhaps
these observations would seem to suggest	these observations suggest
one of the problems	one problem
in only a very small number of cases	occasionally, rarely
in the possible event that	if
An additional piece of evidence that helps to support this hypothesis	Further evidence supporting this hypothesis
In spite of the fact that our knowledge at this point is far from complete	Although our present knowledge is incomplete

WORDY	CONCISE
It is also worth pointing out that	*omit it*
Before concluding, another point is that	*omit it*
It is interesting to note that	*omit it*

■ Avoid repetition.

Some sentences or paragraphs are wordy because the writer has included the same information twice.

WORDY	In Kohmoto's study in 1977, she failed to account for temperature fluctuations (Kohmoto 1977).
CONCISE	Kohmoto (1977) failed to account for temperature fluctuations.

Because the author's name and the publication date are given in the literature citation (Kohmoto 1977), you need not give the same information at the beginning of the sentence. (See also Chapter 4.)

In the next example, the second sentence can be omitted because it merely repeats part of the first sentence using different wording.

> Male fathead minnows who were tending eggs spent the majority of their time rubbing the egg batches with their dorsal pads and preventing other fish from eating the eggs. They did not devote much time to other activities, but instead were chiefly occupied with behavior directed towards the eggs.

■ Use the passive voice sparingly.

In the passive voice, the subject of the sentence *receives* the action, whereas in the active voice it *does* the action.

PASSIVE	Nearly half the seedlings *were eaten* by woodchucks.
ACTIVE	Woodchucks *ate* nearly half the seedlings.

Biological writing leans heavily on the passive voice even though the active voice is more direct, concise, and effective. Why this emphasis on the passive? Compare the following two sentences:

PASSIVE	Skin extract solution *was presented* to the fish through a plastic tube.
ACTIVE	*I presented* skin extract solution to the fish through a plastic tube.

Neither sentence is incorrect; however, the sentence in the passive voice shifts the reader's attention away from the writer and more appropriately to the materials he or she has been testing, observing, collecting, or measuring. This is why the passive voice is particularly common in the Materials and Methods section of biological papers. Shunning the passive altogether may result in prose that is "I-heavy" and monotonous because of too many first-person references. Prudent use of passive constructions gives variety to the text, and in a few situations such constructions may simply be more convenient. The passage below, from the Methods section of a paper by Burger (1974, p. 524), illustrates an effective mix of active and passive voice.

> I used several marking techniques on nests, adults, and juveniles. In 1969, I marked nests with red, blue and white plastic markers tied to cattails. . . . Markers placed on nests were subsequently covered with fresh nest material. Adults were captured with a nest trap . . . and marked with coloured plastic wing tags (Saflag). I marked pairs who were close to the blind by pulling a string attached to a cup of red dye suspended over the nest.

Notice that the author of this passage uses several first-person references (*I* used . . .; *I* marked . . .). Today many biologists are writing "I" instead of impersonal and cumbersome language such as "this investigator" or "the author." This use of the first person makes their prose more direct and concise. It may also reflect a growing realization by biologists that there *is* an "I" in science — that scientific research is inevitably influenced by the personal background, interests, motives, and biases of each researcher.

Thus the passive voice has legitimate uses. Unfortunately, however, many beginning writers rely heavily on the passive because they think it makes their prose more formal, more important, more "fitting" for a scientist. When used habitually, carelessly, or unintentionally, the passive voice results in a wordy and cumbersome style. Overuse of the passive may be one reason why scientific writing has the reputation of being dry, pompous, and boring. Usually the passive voice can easily be converted to the active voice, making sentences shorter and more forceful without any loss of meaning.

Passive	Territory size was found to vary with population density.
Active	Territory size varied with population density.
Passive	From field observations, it was shown that virtually all tagged individuals remained in their original home ranges.
Active	Field observations showed that virtually all tagged individuals remained in their original home ranges.

PASSIVE	Nest destruction was caused primarily by raccoons, particularly late in the incubation period, when greater access to nests was afforded to them by lowered water levels.
ACTIVE	Raccoons caused most nest destruction, particularly late in the incubation period when lowered water levels afforded them greater access to nests.

You will still find the passive voice in most of the scientific literature you read. As you write your own papers, you may feel the need to use the passive voice. There are no firm guidelines about this in biological writing. However, the active voice generally does a much better job. Use passive constructions deliberately and sparingly. For every sentence you put in the passive, ask yourself if you could express it more exactly and concisely in the active voice.

CHAPTER 6

Mechanics and Technicalities

NAMES OF ORGANISMS

The common names of plants and animals have arisen through long traditions of folklore and popular usage. Many reflect some aspect of an organism's appearance, natural history, or relevance to human life: lady's slipper, hedgehog, bedbug, horsehair worm, liverwort. However, common names can also be misleading. Ladybugs, strictly speaking, are not bugs (Hemiptera) but beetles (Coleoptera), and club mosses (Lycopsida) are not true mosses (Musci). Because common names are not universally agreed upon, a single species may be called by different names in different localities. Also, the same common name may be used for two or more taxonomically unrelated organisms.

On the other hand, each species of organism has just one scientific name. It consists of two parts: first, the *genus* to which the organism belongs; and second, the particular *species*. Both generic and specific names are either Latin words or Latinized forms. For example, the scientific name of red maple is *Acer rubrum; Acer* is the genus (to which many different species of maples belong). *Acer rubrum* is the particular species, red maple.

The scientist who first publishes the scientific name of a newly recognized species is regarded as its "author"; his or her last name is placed after the generic and specific names: *Pimephales promelas* Rafinesque. An author's name is often abbreviated: R. for Rafinesque, L. for Linnaeus, Fab. for Fabricius. (However, sources may vary with respect to the abbreviations used for the same author's name.) Sometimes an author's name is put in parentheses, indicating that the species has been put into a different genus from the original. In botany and microbiology but not zoology, the name of the person responsible for this new classification is also added at the end. For example, in the case of *Kickxia spuria* (L.) Dum., the plant was originally called *Linaria spuria* by Linnaeus but was later placed in a different genus, *Kickxia,* by Dumortier (Dum.).

Thus, the scientific names of organisms are determined according to a system of universally accepted rules and conventions. Biologists rely heavily on scientific names; they use common names less frequently. (In fact, most species known to science lack common names.) It is acceptable to refer to organisms by common names as long as you have first given full scientific names along with other taxonomic information, if necessary. Here are some general guidelines for using scientific and common names:

1. The scientific names of species are always italicized (or underlined if you are writing in longhand or with a typewriter or computer). The genus name is capitalized; the species name is not: *Plumularia setacea.* Subspecies or other names below the level of species are italicized but not capitalized: *Plectrophenax nivalis subnivalis.* Author names or their abbreviations are capitalized but not italicized: *Haematosiphon inodorus* (Duges).

2. Give the full scientific name (genus and species) the first time a particular species is referred to in the text. In later references to this species you may abbreviate the genus by its first letter (still capitalized and italicized). The bacterium *Spirillum volutans* thus becomes *S. volutans.* Confusion may arise, however, if you are discussing two or more organisms whose generic names begin with the same letter — for example, *Spirillum volutans* and *Streptococcus salivarius.* In that case, it is safer to spell out the generic names.

3. Generally, the author of a species is specified only the first time the species is mentioned in the paper. (Some biologists include this information in the title, as well.) When you make later references to the species, it is customary to omit the author's name. These conventions apply only to the species in your *own* study. You need not provide such complete taxonomic information when discussing other organisms in other studies — just genus and species names are needed.

4. Do not put an article (*the, a, an*) immediately before the scientific name of a species.

Incorrect	The most common lichen at both sampling sites was the *Lecidea atrata.*
Correct	The most common lichen. . . was *Lecidea atrata.*

5. Do not pluralize scientific names.

Incorrect	Digger bees (*Centris pallidas*) were also observed at the study site.
Correct	Digger bees (*Centris pallida*) were. . .

6. Except in keys (identification guides) or other taxonomic writings, the specific name of an organism must always be preceded by the generic name or its abbreviation.

INCORRECT	The most common species at Hamilton Creek were *Achnanthes minutissima* and *Meridion circulare. Circulare* was also the most common alga at Woods Creek later in the sampling period.
CORRECT	The most common species . . . *Meridion circulare. M. circulare* was also. . . .

The names of genera, however, may be used alone if you are referring collectively to the species in a particular genus.

> Insecticides have been used in an attempt to eradicate *Anopheles* mosquitoes and thus control the spread of malaria.
>
> Some species of *Sargassum* grow in dense mats on the surface of the ocean.

7. Taxonomic groups, or taxa, above the level of genus (family, order, class, phylum, division, and so on) are capitalized but not italicized or underlined.

> The Chilopoda (centipedes) and the Diplopoda (millipedes) are two of six classes in the subphylum Mandibulata.

8. Biologists frequently drop or modify the endings of taxa to make common names for organisms — for example, chironomids from Chironomidae; lycopsids from Lycopsida; dipterans from Diptera; cephalopods from Cephalopoda. Such words, unlike the formal names for groups, are not capitalized.

9. Similarly, other common names of organisms are not usually capitalized except in accordance with specific taxonomic guidelines for certain groups. For example, the common names of American birds have been standardized by the American Ornithologists' Union and *are* capitalized: American Robin, Chipping Sparrow, Barn Swallow. Shortened forms of these proper common names are not capitalized: robin, sparrow, swallow. If a common name contains a word derived from the name for a particular person or place, then that word is generally capitalized even if the rest of the common name is not: English ivy, Queen Anne's lace.

10. Biologists use conventional abbreviations to refer to one or more undesignated species of a particular genus:

> Most nematodes collected from Site 4 were affected by a fungal parasite (*Myzocytium* sp.) [one species].
>
> Francini (1970) found that in certain butterflies (*Colias* spp.) . . . [more than one species].

TIME

Biologists do not report times as a.m. or p.m. Instead, they use a 24-hour time system, numbering the hours of the day consecutively starting at midnight (0000 hours). For example:

7:30 a.m. = 0730 hours
12:00 noon = 1200 hours
9:45 p.m. = 2145 hours
11:53 p.m. = 2353 hours

The word *hours* is often abbreviated:

> Thermoregulatory observations were conducted daily between 0900 and 1600 h throughout July and August, 1987.

Note that *h* is not followed by a period unless it comes at the end of the sentence.

Biological writing also uses a convenient shorthand to specify *photoperiod* (the number of hours of light in a 24-hour period). For example, instead of writing, "Hamsters were exposed to a photoperiod of 12 h," a writer could say, "Hamsters were reared under conditions of 12L:12D," (12 hours of light followed by 12 hours of darkness). Similarly, 14L:10D means 14 hours of light and 10 hours of darkness.

VERB TENSE

Scientific ethics have given rise to the convention of using the *past* tense when reporting your own present findings and the *present* tense when discussing the published work of others. This is because new data are not yet considered established knowledge, whereas the findings of previous studies are treated as part of an existing theoretical framework. Therefore, in a research paper you will need to use both the past and present tenses. Most of the Abstract, Materials and Methods, and Results sections will be in the past tense because you are describing your own work. Much of the Introduction and Discussion will be in the present tense because they include frequent references to published studies. Look, for example, at the following examples.

> The reproductive success of yellow-bellied marmots (*Marmota flaviventris*) is strongly influenced by the availability of food and burrows (Andersen, Armitage, and Hoffmann 1976) [Introduction].

> After inoculation, plants were kept in a high-humidity environment for 100 h [Materials and Methods].

> Limpets occurred at all sampling sites [Results].

> *Diaptomus minutus* was dominant in the zooplankton of Clinton Lake during both years of the study. This species *is* common in many other acidic lakes in the region (Lura and Lura 1985) [Discussion].

When you refer to an author directly, however, you may use the past tense:

> Bauman (1959) found that this bacterium is highly sensitive to pH.

When you refer directly to a table, figure, or a statistical test in your own paper, it is acceptable to use the present tense:

> Table 3 shows that polychaetes were most abundant at depths of 10–16 m.

See the sample review paper at the end of Chapter 1 for other examples of correct verb tense.

PUNCTUATION

Used correctly, punctuation marks help make your writing clear and understandable. Used incorrectly, they may distract, annoy, or confuse the reader. Buy a writer's guide with a detailed section on punctuation and consult it frequently; several good handbooks are listed at the end of this book. The guidelines below address some of the most common punctuation problems in biology writing.

Comma

1. Use a comma to separate introductory material from the rest of the sentence.

> Although egg cases were reared using Hendrickson's (1977) method, none of the eggs hatched.
>
> During the 1987 breeding season, bullfrogs were sexually active from early June to late July.
>
> Nevertheless, these data suggest that mate selection in this species is based primarily on female choice.

A comma may be omitted after a very *short* introductory element that merges smoothly and unambiguously with the rest of the sentence.

> Thus the data in Table 1 are consistent with those of Tables 2 and 3.

2. Use a *pair* of commas to set off insertions or elements that interrupt the flow of a sentence.

> The situation is different, however, on isolated islands with lower species diversity.

> This explanation, first proposed by Hess (1967), is still widely accepted.

3. Use a comma to separate all items in a series, including the last two.

> Unlike mosses, ferns possess true roots, stems, and leaves.

4. Use a comma to separate independent clauses linked by a coordinating conjunction (*and, but, or, for, nor, yet*). (An independent clause is a word group containing a subject and a verb and able to function as a complete sentence.)

> Many studies have been made of feeding preferences in spiders, but few have been done under natural conditions.

Do *not* use a comma on its own to join two independent clauses. Such an error is called a *comma splice*.

> **COMMA SPLICE** Horsetails usually grow in moist habitats, some occur along dry roadsides and railway embankments.

A comma splice also results when you join two independent clauses with a comma followed by a conjunctive adverb (*however, moreover, furthermore, therefore, nevertheless*).

> **COMMA SPLICE** Horsetails usually grow in moist habitats, however some occur along dry roadsides and railway embankments.

There are several ways to correct a comma splice. For example, you can use a semicolon (see p. 119) with or without a conjunctive adverb:

> **REVISION 1** Horsetails usually grow in moist habitats; however, some occur along dry roadsides and railway embankments.

> **REVISION 2** Horsetails usually grow in moist habitats; some occur along dry roadsides and railway embankments.

A comma splice can also be corrected by adding a coordinating conjunction after the comma:

> **REVISION 3** Horsetails usually grow in moist habitats, but some occur along dry roadsides and railway embankments.

Alternatively, you can make one of the clauses a dependent clause, and separate the two clauses by a comma. (A dependent clause lacks either a subject or a verb and cannot stand alone as a complete sentence.)

REVISION 4 Although horsetails usually grow in moist habitats, some occur along dry roadsides and railway embankments.

Finally, you can make two sentences separated by a period:

REVISION 5 Horsetails usually grow in moist habitats. Some occur along dry roadsides and railway embankments.

Semicolon

1. Use a semicolon to join two independent clauses *not* linked by a coordinating conjunction (*and, but, or, for, nor, yet*).

> Bacteria reproduce very rapidly; many species can divide once every 20 minutes under favorable conditions.

This statement could also have been written as two complete sentences separated by a period, but the semicolon suggests a closer relationship between the two ideas.

2. Use a semicolon to connect two independent clauses joined by a conjunctive adverb (*however, moreover, nevertheless, furthermore*).

> These data are consistent with those of Allen (1980); moreover, they suggest that pH may be a more important influence than previously believed.

3. Use a semicolon to clarify the meaning of a series of items containing internal commas.

> Films were made of courtship and mating behavior; aggressive interactions between males, including chasing, butting, and biting; and parental behavior by females, particularly fanning and egg retrieving.

Colon

1. Use a colon to introduce items in a series.

> Birds are distinguished by the following features: a four-chambered heart, feathers, light bones, and air sacs.

However, do *not* use a colon before a series unless the colon follows an independent clause.

> Dandelions are in the same family as: daisies, chrysanthemums, sunflowers, and hawkweeds.

> The most common insects at Site A were: mayflies, dragonflies, damselflies, and butterflies.

In both these examples, the colon should be deleted.

2. Use a colon between two independent clauses when the second clause explains or clarifies the first one.

> The Cranston Lake study site was most suitable for incubation studies: nest sites were abundant, and brooding birds were rarely disturbed by road traffic.

3. Use a colon to formally introduce a direct quotation. (See pp. 54–55.)

Dash

The dash is used to separate material abruptly from the rest of the sentence for clarity, emphasis, or explanation. (On your typewriter, you would use two hyphens, one after the other, with no space on either side.)

> Insects are distinguished by three body regions — head, thorax, and abdomen — and three pairs of legs.

Many writers use the dash when a comma, indicating just a slight pause, would be less distracting.

Distracting	Eggs with minor cracks and an intact inner membrane hatched normally — but severely cracked eggs failed to develop and soon rotted.
Revised	Eggs with minor cracks and an intact inner membrane hatched normally, but severely cracked eggs failed to develop and soon rotted.

When overused, dashes make your writing seem informal and careless, suggesting that you do not know how to use other types of punctuation. Use dashes sparingly, if at all, in biological writing.

Parentheses

1. Use parentheses to insert explanatory or supplemental information into a sentence.

> The nests at Leland Pond were unusually large (at least 30 cm in diameter), but only one contained any eggs.
>
> Juvenile monkeys raised in groups of three gained weight more rapidly than those raised in isolation (Fig. 2).

Like dashes, parentheses should be used sparingly because they interrupt the flow of your writing. Commas often work just as well as parentheses and are less distracting.

2. Use parentheses with a series of items introduced by letters or numbers, especially when you are listing many items or several long items.

> Dragonflies are good subjects for studies of territorial behavior for

these reasons: (1) many are easily identified and widely distributed, (2) most can be observed easily under natural conditions, and (3) much is known about the anatomy and life history of many species.

Quotation Marks, Brackets, Ellipsis

See pages 54–55 for uses of these punctuation marks to present quoted material.

SYMBOLS AND ABBREVIATIONS

Following are some symbols and abbreviations commonly used in biology. Note that many are not followed by a period (unless they appear at the end of a sentence).

Term/Unit of Measurement	Symbol/Abbreviation
ångström	Å
approximately	c. *or* ≈
calorie	cal
centimeter	cm
cubic centimeter	cm^3
cubic meter	m^3
cubic millimeter	mm^3
day	d
degree Celsius	°C
degree Fahrenheit	°F
diameter	d, Diam
degrees of freedom	df
east	E
et alii (Latin: and others)	*et al.* *or* et al.
et cetera (Latin: and others)	etc.
exempli gratia (Latin: for example)	*e.g.* *or* e.g.
female	♀
figure, figures	Fig., Figs.
foot-candle	fc *or* ft-c
gram	g
greater than	$>$
hectare	ha
height	ht
hour	h, hr
ibidem (Latin: in the same place)	*ibid.*
id est (Latin: that is)	*i.e.* *or* i.e.
joule	J
kelvin	K

Term/Unit of Measurement	Symbol/Abbreviation
kilocalorie	Kcal
kilogram	kg
kilometer	km
latitude	lat.
less than	$<$
liter	l *or* L *or* liter *to avoid confusing with the numeral* 1
logarithm (base 10)	log
logarithm (base *e*)	ln
longitude	long.
male	♂
maximum	max
mean	$\bar{x}$
meter, metre	m
microgram	μg
microliter	μl
micrometer (micron)	μm
milliliter	ml
millimeter	mm
minimum	min
minute (time)	min
molar (concentration)	*M*
mole	mol
month	mo
nanometer	nm
north	N
not (statistically) significant	NS
number	No.
number (sample size)	*N or n*
parts per million	ppm
percent	%
plus or minus	±
probability	*P or p*
second (time)	s
south	S
species (singular)	sp.
species (plural)	spp.
square centimeter	cm^2
square meter	m^2
square millimeter	mm^2
standard deviation	SD
standard error	SE

TERM/UNIT OF MEASUREMENT	SYMBOL/ABBREVIATION
standard temperature and pressure	STP
versus	vs
volume	vol
volt	V
watt	W
week	wk
weight	wt
west	W
year	yr

MANUSCRIPT FORMAT

After making final revisions on your paper, you may feel that all the work is over. It is — almost. You still need to produce a neat, clean manuscript. A sloppily prepared paper, one with page numbers missing, margins askew, and the type barely visible, will not show off your prose to good advantage. Such details affect the reader's overall impression of your work.

Below are some general guidelines for preparing student papers in biology. These are generally similar to the criteria specified by most biological journals. Your instructor may have his or her own specifications; if not, following these conventions will help you produce a professional-looking manuscript.

Paper, Margins, and Spacing

Use 8½-by-11-inch white bond paper (not erasable bond or onionskin), a dark ribbon on your typewriter or printer, and a standard typeface (not script type or all capitals). Type on only one side of the paper. Leave two spaces after each period and a single space after each comma. Indent each paragraph five spaces. Double-space the entire manuscript, including the Abstract, the Literature Cited section (between adjacent references as well as between the lines of a single reference), table titles, figure legends, and any indented quotations in the text. Leave margins of 1–1½ inches on all sides of the page.

Pagination

Number pages consecutively beginning with the title page (which does not actually carry a number, but is still counted). Use only arabic numerals (1, 2, 3, . . .) and put the number in the upper right-hand corner of each page. Writers submitting a research paper for publication often begin each major section of the paper (Abstract, Introduction, Materials and Methods, and so on) on a separate page. This is not necessary for a student paper, although you may find that separating the sections makes a long paper easier to follow.

Put each table or figure on a separate page. Type the title of each table on the same page as the table. Journal specifications call for all figure legends to be typed together on a separate page. Tables, figures, and figure legends are then placed at the end of the Literature Cited section in that order. If your instructor prefers that tables and figures be incorporated with the text, put each table or figure after the page on which it is first mentioned and number all pages. In this case, each figure will have to be accompanied by its legend, typed either on the same or a separate page.

If the paper was typed by a computer printer, separate the individual pages and arrange them in their proper order. A manuscript submitted as one continuous sheet of computer paper is an inexcusable sign of sloppiness.

Title Page

An acceptable format for a student paper calls for the following information on the title page: title of the paper, author(s), course information, and date. Center each line on the page, and capitalize the first letter of all important words.

Proofreading

Typographical errors, misspelled words, missing commas or periods, irregular spacing, and other minor errors distract the reader and undermine your authority as writer. The aim of proofreading is to eliminate such mistakes from the final draft of the manuscript.

It is not easy to proofread well. By the time you have worked through several drafts of the paper you are no longer as observant as you might be. You will tend to read what you *think* is there, not what is actually on the page. Therefore you must force yourself to look at every word, space, line, punctuation mark, and number, including all material in figures and tables. It is wise to proofread the whole manuscript more than once over a period of several hours or days. You may miss some errors the first time but notice them during the second or third proofreading, when you have achieved more "distance" from the material. Ask a friend to look over the paper, as well.

If you write using a word processor, learn to use features that speed up revising and proofreading (pp. 93–95).

Finally, use a staple or a paper clip to fasten everything together; a separate folder is not necessary. In case your instructor misplaces the paper, keep a copy of the original.

CHAPTER 7

Other Forms of Biological Writing

CLASS NOTES

Over the years, you may have developed your own strategies for taking notes during lectures and from course readings. By now you know that good class notes can be invaluable for self-teaching. The following suggestions will help you use writing more effectively to study biology and many other subjects.

1. During lectures, do not feel you have to record every word. If you write continuously, you will not be giving full attention to what the speaker is saying. Your notes may be voluminous, but they may not make sense because you won't be thinking about what you are writing. Taking notes nonstop also inhibits your ability to ask questions about what you do not understand and to contribute to class discussions.

Good lecture notes are a *selective* record. They reflect the writer's attempts to summarize and synthesize the most important points, to reproduce the logic and continuity of the speaker's discussion, and to make links to material from previous class meetings. Many of your notes may be abbreviated versions of the lecturer's own sentences, but others should be in your own words, particularly when you need to bridge the gap between one topic and another.

2. Keep up with assigned readings. Instructors write their lectures assuming students have acquired certain background knowledge by reading the textbook or other course materials. If you come to class with that knowledge, you are more likely to understand the lecture and to take

coherent notes. Moreover, you will be less likely to waste time and energy writing down information covered adequately elsewhere.

3. Revise your notes soon after the lecture, preferably the same day. You probably use various shortcuts to take notes quickly, such as writing in sentence fragments and abbreviating words. Your writing may make sense at the time but be hard to decipher later. For this reason, it pays to go over each day's notes while the material is still fresh in your mind. Write out in full any shortened sentences or words that may confuse you later. Clarify difficult concepts by adding additional comments in the margins. Check that diagrams are accurately labeled and integrated with the rest of your notes. Number related items and use underlining, asterisks, or other labels to call attention to the most important ideas. Reworking your notes after each lecture may seem time-consuming, but it will pay off later when you are studying for examinations. Moreover, it reinforces your understanding of the material.

4. Make reading an active, not a passive, experience. Annotate your textbooks by writing brief notes, comments, and questions in the margins. Make difficult passages comprehensible by paraphrasing or summarizing them in your own words (see pp. 45–46). Underline or highlight important sentences — but do so sparingly and selectively. *First* read a passage all the way through; then go over it again and make notations that will help you make sense of the material later. Avoid highlighting line for line *as* you are reading; doing so usually does not allow you to make meaningful decisions about the relative importance of related sentences.

Don't ignore summaries or study questions at the ends of chapters. These reflect the author's insights about the most significant material.

5. Laboratory and field notes benefit enormously from advance preparation. Read the laboratory manual *before* you come to class, and understand why you are doing each experiment. Jot down observations and questions as you work; do not trust your memory. Label diagrams carefully; note magnification for microscope drawings. Reread your notes later, adding additional comments and explanations; such material will be invaluable when you are preparing the laboratory report. (See also pp. 3–5 on preparing data sheets.)

6. Do not study for tests simply by rereading your notes and the textbook. Use writing to become more actively involved with the subject. An excellent study strategy is to create your *own* "textbook" by integrating lecture notes, course readings, and material from laboratory or field sessions. After all, your instructor sees all these components as related to one another; you should, too. Next, compile lists of key terms and concepts; after each, write out its definition or significance. Finally, devise your own examination questions, both essay and short-answer types, and answer them with "ideal" answers. It often helps to group related items: three factors that influence a particular phenomenon; similarities between two groups of organisms;

six characteristics of a particular family of animals, and so on. Even when studying for short-answer tests, focus on central issues and generalities. Biology is not just a collection of isolated facts; much more important are the underlying ideas.

ESSAY QUESTIONS

Clear, coherent writing is as important in essay examinations as it is in other forms of biological writing. Awkward, wordy, or ungrammatical prose will suggest (probably correctly) that your thinking is also sloppy and unrefined. The problem, of course, is that time constraints during examinations usually do not allow much time for reworking and refining your writing. For this reason, good essays are the result of well-focused preparation beforehand, followed by careful use of time during the examination. The following strategies may help you.

1. Read through the whole examination before you begin any question. Check whether you must answer all questions or if you have some choice. Start with the easiest sections of the test; leave difficult ones for later. Budget your time carefully; give yourself a time limit for each question. Leave a few minutes at the end to review your answers and to add additional material, if necessary. Check again that you have answered or attempted all required questions.
2. Address the question *asked.* This may sound obvious; however, many students do not read the question carefully. Focus on key words in the question; for example, define, list, compare or contrast, evaluate, analyze, describe, discuss. Check if examples are required in addition to more general statements. Pay attention to length requirements for answers and to opportunities for choice within questions.
3. Before you start to write, make a brief plan or jot down a list of important points or concepts. It is easy to forget crucial points when you are working under pressure. Taking a minute or two to organize your thoughts will help you write a more complete answer.
4. Develop a clear thesis. Essays requiring you to integrate, analyze, or evaluate material can easily lose focus. Do not ramble; organize your essay around a clearly stated central idea. State this in a topic sentence (see p. 97); then build the rest of the essay around your main point.
5. Support general statements with specific evidence or examples. Many essays are inadequate because the writer conveys only a superficial knowledge of the material. Thorough understanding of a subject involves the ability to show how broad statements rest on a foundation of supporting details.
6. Stick to the point. You will not get a higher grade for adding irrelevant information; moreover, such material obscures the good qualities of your essay.

7. Good essays on examinations are usually the product of much advance preparation. If your studying has been thorough and well focused, you will have already thought of many possible essay questions yourself — perhaps some of the same ones the instructor has devised. In other words, you will have already tried to make connections between topics, to identify central issues, and to see the course material in new ways. All these are purposes of essay questions.

ORAL PRESENTATIONS

Many biologists participate in professional conferences or symposia that offer a forum for exchanging ideas, conveying information about new research methods, and reporting research in progress. Conference proceedings are often available in printed form; however, oral exchange of information is still the primary means of communication.

As a biology student, you may be required to prepare an oral presentation of your own — perhaps a summary of a published paper, a report of your own research, or a review of the literature on a particular topic. Although the final product will not be presented in written form, you will still need to use writing to organize your talk. Actually, there are many similarities between writing a paper and planning an oral presentation. Both activities require you to consider your audience and your purpose and to convey information clearly, accurately, and logically. Both also force you to examine your own understanding of the material and to use writing as a means of clarifying your thoughts.

Following are some practical suggestions for giving successful oral presentations.

1. Write out the entire talk beforehand. Even if you are an accomplished speaker, putting everything down in writing will make your presentation more organized and coherent and will lessen the chance that you'll forget an important point. Plotting out the talk sentence by sentence also gives you time to plan your words carefully and search for the most effective ways of explaining difficult ideas. After you rough out the first draft, revise the manuscript carefully, looking for sections that are poorly worded or likely to confuse the audience, and for abrupt jumps from one topic to another.

2. Never *read* a prepared talk word for word. Doing so will probably annoy your instructor, who will assume you are not comfortable enough with the material to abandon your notes. It will also distance you from your listeners, who will rapidly lose interest in your formal recitation. Instead, use the written version of the talk to make a brief list of key points or concepts; these can be put on a single sheet of paper or on file cards arranged sequentially. As you speak, use these key points to jog

your memory and keep you on track. Once you become thoroughly familiar with the talk in its written form, you will probably remember your most effective sentences and phrases word for word and will be able to say these naturally as if they just occurred to you. An effective talk — one that really engages the audience — strikes the proper balance between carefully structured wording, worked out in advance, and a spontaneous, informal delivery.

3. Observe your time limit. Practice speaking in front of a friend or rehearse the talk mentally, timing the entire presentation from start to finish. Determine what you should be discussing halfway through the allotted time period, and note this point both on the final draft of the talk and on your speaking notes.

4. If you are nervous you may speak more rapidly than usual, especially as you approach the end of the talk. Remember that listeners need time to digest everything they hear. If you confuse them at any point, they may stay confused for the rest of the talk; unlike readers of written text, they cannot easily go back and review difficult sentences. It helps to pause briefly after important points and to repeat difficult material in slightly different wording. If your format permits, you can also invite questions at potentially confusing places in the presentation.

5. Establish eye contact with the audience. Doing so will make you feel more relaxed when speaking and will make your audience more receptive.

6. How you organize the presentation will depend on the subject and your objective. Planning may be easiest if you think of the talk as having a distinct beginning, middle, and end. You might start by briefly reviewing general background material, then narrowing down to your specific objectives. Then introduce major concepts and findings, adding just enough supporting information and examples to develop your points. Listeners can take in only so much at one time; many people will lose track of the central argument if you bombard them with too many details. Use straightforward language; avoid jargon (see pp. 106–108). Explain any terms likely to be unfamiliar, but keep the number of such terms to a minimum. Save time at the end to summarize the most important points, offer conclusions, and discuss broader aspects of the topic.

7. Use visual aids, if possible. The blackboard is an invaluable tool even for short presentations. Take advantage of it as you talk to write down unfamiliar terms or important statistics and to draw simple tables and graphs. Your writing should be large enough to be seen clearly from the back of the room. Graphs should be simple to read and clearly labeled. Do not clutter the board with disorganized scrawls, and do not inadvertently erase information you must refer to frequently.

Material that is relatively time-consuming to draw should be put on the blackboard *before* your talk; otherwise you'll waste time writing while the audience waits impatiently or struggles to copy what you have written. It may make more sense to distribute such material as a supplementary handout that the audience can look at as you talk. Handouts are also useful for listing key terms and definitions, important points to be covered, or useful references. However, do not make handouts too detailed or extensive, or people will spend their time reading instead of listening to you.

If you are presenting an original research project, you may wish to show slides illustrating various aspects of the study, such as the field site, organisms studied, experimental apparatus, and quantitative data in the form of tables or graphs. However, to be effective slides must be of good quality. Do not make the common mistake, seen even at professional meetings, of showing poorly exposed, blurry, or otherwise unsuitable slides accompanied by repeated apologies ("Sorry these slides are so dark, but . . ."). Such slides are not worth using at all. If you are not a competent photographer, check the facilities of the audiovisual department at your institution, or think of an alternate method for illustrating your talk.

If you do have sharp, uncluttered, well-exposed slides, spend time planning how you will integrate these with the "text" of your talk. As in a written paper, each illustration should illustrate or support a particular statement. Write out beforehand what you plan to say about each slide. Omit any that serve no clear purpose. Write down on your lecture notes when each slide should be shown. Where possible, try to show slides in one or more groups; otherwise, you will distract the audience by repeatedly flicking the lights on and off. Set up the projector and screen in advance, locate a pointer, and check that all slides are the right way up and in the proper order.

Like blackboard illustrations, slides depicting numerical data should be simple enough for the audience to grasp quickly. Do not pack them with more information than people can absorb in a minute or two. Axes of graphs should be clearly labeled, and all words and numbers should be large enough for viewers to read on their own. Remember that slides showing data should generally stress concepts, not particulars. Use them carefully and sparingly to make your points.

Overhead transparencies of figures and tables are also effective lecture illustrations. Consider using blank transparent sheets to jot down terms or make quick drawings as you talk. An advantage of overhead projection is that it allows the speaker to leave the room lights on and thus to interact more directly with the audience.

8. Be prepared for questions. You cannot predict everything you will be asked, but you probably can anticipate some of the questions. Write these out beforehand and prepare brief, concise answers. If you are asked a question for which you are unprepared, do not try to bluff your way through a reply. It is far better to say that you don't know the answer. If you have given a thoughtful, well-organized talk, listeners will already be convinced that you know your subject. They will not expect you to know everything.

Additional Readings

General Writing Handbooks

Every writer should have at least one general writing guide. The following handbooks cover the essentials of grammar, punctuation, and word choice as well as the crafting of sentences, paragraphs, and whole essays. All are organized as self-help manuals presenting easy-to-follow guidelines applicable to a wide variety of writing situations.

Baker, S. 1985. The practical stylist. 6th ed. Harper and Row, New York.

Dornan, E. A., and C. W. Dawe. 1987. The brief English handbook. 2nd ed. Little, Brown and Co., Boston.

Fowler, H. R. 1986. The Little, Brown handbook. 3rd ed. Little, Brown and Co., Boston.

Glenn-Leggett, C., C. D. Mead, and M. G. Kramer. 1985. Prentice-Hall handbook for writers. 9th ed. Prentice-Hall, Englewood Cliffs, New Jersey.

Hacker, D. 1988. Rules for writers. 2nd ed. Bedford Books of St. Martin's Press, New York.

Hairston, M. C. 1981. Successful writing. 2nd ed. W. W. Norton, New York.

This book emphasizes the writing process and includes a thorough and helpful discussion of the act of revising.

Perrin, R. 1987. The Beacon handbook. Houghton Mifflin Co., Boston.

Strunk, W., Jr., and E. B. White. 1979. The elements of style. 3rd ed. Macmillan, New York.

Dictionaries

Allaby, M. 1983. A dictionary of the environment. 2nd ed. New York University Press, New York.

The American heritage dictionary: second college edition. 1982. Houghton Mifflin Co., Boston.

Campbell, B., and E. Lack. 1985. A dictionary of birds. Buteo Books, Vermillion, South Dakota.

Dorland's illustrated medical dictionary. 24th ed. 1965. W. B. Saunders Co., Philadelphia.

Dox, I., B. J. Melloni, and G. M. Eisner. 1979. Melloni's illustrated medical dictionary. Williams and Wilkins, Baltimore.

Hanson, H. H. 1962. Dictionary of ecology. Philosophical Library, New York.

King, R. C., and W. D. Stansfield. 1985. A dictionary of genetics. 3rd ed. Oxford University Press, New York.

Lewis, W. H. 1977. Ecology field glossary: a naturalist's vocabulary. Greenwood, Westport, Connecticut.

Little, R. J., and C. E. Jones. 1980. A dictionary of botany. Van Nostrand Reinhold Co., New York.

The McGraw-Hill dictionary of scientific and technical terms. 2nd ed. 1978. McGraw-Hill, New York.

The McGraw-Hill dictionary of the life sciences. 1976. McGraw-Hill, New York.

Swartz, D. 1971. Collegiate dictionary of botany. The Ronald Press Co., New York.

Urdang dictionary of current medical terms for health science professionals. 1981. John Wiley and Sons, New York.

Guides to Scientific Writing

The references that follow are particularly useful for more experienced scientific writers, including professional biologists, and for undergraduate and graduate students preparing a paper for a technical journal.

CBE style manual committee. 1983. CBE style manual: a guide for authors, editors, and publishers in the biological sciences. 5th ed. rev. and expanded. Council of Biology Editors, Inc., Bethesda, Maryland.

This standard reference for biologists in all fields can be found in the reference section of the library or can be ordered from the Council of Biology Editors, Inc., 9650 Rockville Pike, Bethesda, MD 20814.

The Chicago manual of style. 13th ed. 1982. University of Chicago Press, Chicago.

A widely used reference for writers in many academic fields. Covers a wide range of topics, including documentation methods, tables and figures, use of quotations, and many aspects of writing mechanics and style. Useful as a supplement to guides specifically devoted to scientific writing.

A compilation of journal instructions to authors. 1979. U. S. Dept. of Health, Education, and Welfare. NIH Pub. No. 79–1991.

This publication includes guidelines for writing and submitting manuscripts to numerous journals, particularly those related to health and medicine.

Day, R. S. 1983. How to write and publish a scientific paper. 2nd ed. ISI Press, Philadelphia, 1983.

Written with style and wit, this short, readable book gives much practical advice about preparing a paper for publication. It also includes chapters on preparing book reviews, conference reports, theses, and oral presentations.

Huth, E. J. 1982. How to write and publish papers in the medical sciences. ISI Press, Philadelphia.

A detailed, thorough guide to writing medical papers, with a useful annotated bibliography.

Scientific writing for graduate students: a manual on the teaching of scientific writing. Committee on Graduate Training in Scientific Writing. F. Peter Woodford, ed. Council of Biology Editors, Inc., Bethesda, Maryland, 1986.

Although written primarily for *instructors* of scientific writing, this manual is also useful for both the beginning and advanced student. The book contains thorough coverage of the research paper as well as sections on tables and figures, research proposals, theses, and oral presentations. It can be ordered from the publishers of the *CBE Style Manual* (see p. 132).

Smith, R. C., W. M. Reid, and A. E. Luchsinger. 1980. Smith's guide to the literature of the life sciences. 9th ed. Burgess Pub. Co., Minneapolis.

A detailed guide to the biological literature, including textbooks, dictionaries, review papers, indexes, abstracts, and primary sources. Also discusses the writing of theses, research and review papers, research proposals, and oral presentations.

Word Processing

Hult, C., and J. Harris. 1987. A writer's introduction to word processing. Wadsworth, Inc., Belmont, California.

A readable and practical guide to using word processing in your own writing. It assumes no background in the use of computers.

Statistics

The following texts are good introductions for beginners:

Baily, N. T. J. 1981. Statistical methods in biology. 2nd ed. John Wiley and Sons, New York.

Lewis, A. E. 1984. Biostatistics. 2nd ed. Van Nostrand Reinhold Co., New York.

The texts below, aimed at more advanced readers, give a comprehensive, thorough treatment of many statistical topics.

Sokal, R. R., and F. J. Rohlf. 1981. Biometry. 2nd ed. Freeman and Co., San Francisco.

Zar, J. H. 1984. Biostatistical analysis. 2nd ed. Prentice-Hall, Englewood Cliffs, New Jersey.

Figures and Tables

Cleveland, W. S. 1985. The elements of graphing data. Wadsworth Advanced Books and Software, Monterey, California.

For more advanced readers. Covers a wide variety of graphs, with numerous "good" and "bad" illustrations from the scientific and technical literature.

Selby, P. H. 1976. Interpreting graphs and tables. John Wiley and Sons, New York.

Designed as a self-teaching manual, this short guide is aimed at beginners with little knowledge of the subject.

References Cited

Bem, S. L. 1981. Gender schema theory: a cognitive account of sex typing. Psychol. Rev. 88:354–364.

Burger, J. 1974. Breeding adaptations of Franklin's gull (*Larus pipixcan*) to a marsh habitat. Anim. Behav. 22:521–567.

Dawkins, R. 1976. The selfish gene. Oxford University Press, New York.

Eiseley, L. 1958. Darwin's century: Evolution and the men who discovered it. Doubleday, New York.

Ginsberg, H. J., W. G. Braud, and R. D. Taylor. 1974. Inhibition of distress vocalizations in the open field as a function of heightened fear or arousal in domestic fowl. Anim. Behav. 22:745–749.

Hepler, P. K., and R. O. Wayne. 1985. Calcium and plant development. Ann. Rev. Plant Physiol. 36:397–439.

Kohmoto, K., I. D. Kahn, Y. Renbutsu, T. Taniguchi, and S. Nishimura. 1976. Multiple host-specific toxins of *Alternaria mali* and their effect on the permeability of host cells. Physiol. Plant Pathol. 8:141–153.

McMillan, V., and R. J. F. Smith. 1974. Agonistic and reproductive behaviour of the fathead minnow (*Pimephales promelas* Rafinesque). Z. Tierpsychol. 34:25–58.

Morris, D. 1967. The naked ape. Bantam Books. Toronto, Canada.

Nagel, M., N. Oshima, and R. Fujii. 1986. A comparative study of melanin-concentrating hormone (MCH) action on teleost melanophores. Biol. Bull. 171:360–370.

Raven, P. H., and G. B. Johnson. 1986. Biology. Times Mirror/Mosby College Publishing, St. Louis, Missouri

Young, C. M., P. G. Greenwood, and C. J. Powell. 1986. The ecological role of defensive secretions in the intertidal pulmonate (*Onchidella borealis*). Biol. Bull. 171:391–404.

Index

Symbols and Abbreviations Used in Biology

TERM/UNIT OF MEASUREMENT	SYMBOL/ ABBREVIATION
ångström	Å
approximately	c. *or* $\approx$
calorie	cal
centimeter	cm
cubic centimeter	cm^3
cubic meter	m^3
cubic millimeter	mm^3
day	d
degree Celsius	°C
degree Fahrenheit	°F
diameter	d, Diam
degrees of freedom	df
east	E
et alii (Latin: and others)	*et al. or* et al.
et cetera (Latin: and others)	etc.
exempli gratia (Latin: for example)	*e.g. or* e.g.
female	♀
figure, figures	Fig., Figs.
foot-candle	fc *or* ft-c
gram	g
greater than	$>$
hectare	ha
height	ht
hour	h, hr
ibidem (Latin: in the same place)	*ibid.*
id est (Latin: that is)	*i.e. or* i.e.
joule	J
kelvin	K
kilocalorie	Kcal
kilogram	kg
kilometer	km
latitude	lat.
less than	$<$
liter	l *or* L *or* liter *to avoid confusing with the numeral 1*
logarithm (base 10)	log
logarithm (base *e*)	ln
longitude	long.
male	♂